The Determination of Hydroxyl Groups

# THE ANALYSIS OF ORGANIC MATERIALS

An International Series of Monographs

*edited by*

R. BELCHER and D. M. W. ANDERSON

1. S. VEIBEL: The Determination of Hydroxyl Groups. 1972

# The Determination of Hydroxyl Groups

STIG VEIBEL

*Emeritus Professor*
*Technical University of Denmark*

1972

Academic Press
London and New York

ACADEMIC PRESS INC. (LONDON) LTD.
24/28 Oval Road,
London NW1

*United States Edition published by*
ACADEMIC PRESS INC.
111 Fifth Avenue
New York, New York 10003



Library of Congress Catalog Card Number: 72-84887
ISBN: 0-12-715250-4

PRINTED IN GREAT BRITAIN BY
PAGE BROS (NORWICH) LTD., NORWICH

# PREFACE

THE contents of this volume are based on the experiences gained during more than 40 years of teaching organic chemistry, including organic analytical chemistry, at the University of Copenhagen and Technical University of Denmark. The volume is divided into three parts: (A) Detection of the Hydroxyl Group; (B) Characterization of Hydroxy Compounds; and (C) Estimation of Hydroxy Compounds. For substances where the presence of a hydroxy group and its character (aliphatic, aromatic, etc.) is known, the first two sections are, of course, unnecessary or of minor importance.

In many instances, however, the analytical chemist has first of all to find out which functional group is the most convenient for the estimation of an unknown substance. Part A should provide guidance when the presence of a hydroxy group, and also its character, have to be established.

In other instances, a convenient method for the direct estimation of the hydroxy group may not be available and it may then be necessary to prepare a suitable derivative of the hydroxy compound on which to base the determination. Part B describes methods for the preparation of such derivatives.

The main section is, however, Part C—Estimation of Hydroxy Compounds. For pure hydroxy compounds it is often more convenient to calculate the equivalent weight, i.e. the weight of the compound containing 17 g of OH. It may be more directly illustrative to know that 17 g of OH are contained in 88 g of a compound than to be informed that the compound contains 19·3 % OH. In determining the content of a hydroxy compound in a mixture, however, it is preferable to calculate the latter value, from which the content of a particular hydroxy compound is easily calculated when its molecular weight is known.

Many more methods than those mentioned in this volume have been suggested. It is my hope that the selection presented will prove appropriate for the majority of compounds or mixtures submitted to the analytical chemist for determination as hydroxy compounds, although more refined methods may well have to be found (or elaborated) for special cases.

S. V.

# CONTENTS

## PART C. ESTIMATION OF HYDROXY COMPOUNDS

# INTRODUCTION

THE hydroxyl function is contained in a great variety of organic compounds, either as the sole functional group (as in alcohols, phenols, and heterocyclic hydroxy compounds), or together with a different functional group (as in hydroxy aldehydes, hydroxy acids or hydroxy amines), or as part of a composite group (as in carboxylic acids). The subject of this monograph is the determination of the first-mentioned type of hydroxy compounds. Although methods for the determination of some of the other types will be mentioned, the determination of such compounds is usually more conveniently based on the other functional group present, and will therefore be dealt with in other monographs of this series. Compounds in which the hydroxyl function is part of a composite group will not be discussed here.

The determination of compounds containing a functional group may be based either on a reaction of the functional group itself or on a reaction of a derivative prepared for the characterization of the compound under examination. The complete process of determination therefore usually involves three parts: (A) Detection of the presence of one or more hydroxyl groups by appropriate reactions; (B) Preparation of derivatives suitable for the characterization of the hydroxy compound; and (C) Estimation of the equivalent weight of the compound (or the amount of the compound present in a mixture) by methods based on the hydroxyl group itself or on a derivative. Reactions useful for each of these three parts will be described in the following corresponding parts.

# Part A

# Detection of the Hydroxyl Group

CHAPTER 1

# ALIPHATICALLY BOUND HYDROXYL

## 1. Non-specific Reactions

Most non-specific reactions for the hydroxyl group are so general that they are valid not only for the compounds considered here but also for compounds in which hydroxyl is part of a composite group or for compounds containing sufficiently activated hydrogen, bound directly to carbon or nitrogen. Such reactions involve the evolution of hydrogen on treatment with sodium, the evolution of hydrogen chloride on treatment with phosphorus pentachloride, or the evolution of methane on treatment with methyl magnesium halide. In all these reactions, a trace of water in the compound will result in a positive reaction for hydroxyl.

Not quite so general are some colour reactions, but they do not permit a distinction between different types of aliphatically bound hydroxyl, and at least two of these reactions should give a positive result before the presence of an aliphatically bound hydroxyl group is regarded as established.

### (*a*) *Diazobenzenesulphonic Acid*

Alcohols give a colour reaction with an alkaline solution of diazobenzenesulphonic acid. The colour is maroon to deep red for alcohols readily soluble in water, pink for those that are less soluble. The procedure is as follows:(1)

To a mixture of 1 ml of 0·7% aqueous sodium nitrite and 4 ml of 0·5% solution of sulphanilic acid in N hydrochloric acid are added a few drops or crystals of the compound under examination, followed by an excess of dilute sodium hydroxide solution. The mixture is then heated (but not boiled) for 1–2 min. The colour develops slowly during heating and its intensity may increase during the first 5–10 min after heating is finished.

The most serious limitation of this test is that it is too sensitive. Thus, most phenols will couple with the diazotized sulphanilic acid to form a deeply coloured azo compound. Many aromatic amines will also couple, but both coupling reactions take place instantaneously and are easily distinguished from the slowly developed colour that is characteristic of aliphatic hydroxy compounds.

Another source of error is that some substances, e.g. esters and aliphatic halogen compounds, may be hydrolyzed to hydroxy compounds when treated with hot sodium hydroxide, and a feeble reaction which develops only after heating for some time cannot, therefore, be taken as evidence for the presence of a hydroxyl group in the unchanged compound.

Furthermore, many carbonyl compounds will be sufficiently enolized in alkaline solution to give a feeble reaction for the presence of a hydroxyl group.

Finally, care must be taken to avoid using an excess of sodium nitrite in preparing the diazonium solution since primary aliphatic amines, if present, will be converted into alcohols by the nitrous acid.

### (*b*) *Iron(III) Chloride*

Most hydroxy compounds will give a yellow colour reaction with iron(III) chloride. The intensity of the colour is, however, often so feeble that it is necessary to compare (a) the colour of the solution prepared by adding one or two drops of diluted iron(III) chloride solution to 5 ml of an aqueous solution of the substance to be tested with (b) the colour of a blank prepared by adding one or two drops of iron(III) chloride solution to 5 ml water. If the colour in tube (a) is more intense than that in (b), the test is regarded as positive. It must be remembered, however, that colour reactions with iron(III) ions are not at all specific for aliphatic hydroxy compounds. On the addition of iron(III) chloride many carboxylic acids give a yellow colour, and most phenols and enols will give an intense colour, although not usually yellow. Many amino acids will also give a colour reaction, usually reddish, with iron(III) chloride, and many aromatic amines are oxidized to dark-coloured substances by this reagent.

### (*c*) *8-Hydroxyquinoline Vanadate(V)*

Colour reaction with 8-hydroxyquinoline vanadate, vanadium(V) oxine complex. The 8-hydroxyquinoline salt of vanadic acid is a black complex compound, very sparingly soluble in most solvents. The saturated solution in tetrachloroethane is black;[(2)] in acetic acid it is blue.[(3, 4)] When some drops of an alcohol are added to some ml of the reagent solution, a colour change from black or blue to purple or red takes place within 15 min. Blair and Pantony[(2)] record the following colour changes:

| | |
|---|---|
| Monohydric alcohols | black → red |
| Thiols | black → green or yellow |
| Amines | black → lemon yellow |
| Phenols | black → blue |
| Acetic acid | black → blue (cf. refs. 4, 5). |

The reaction usually fails with polyhydric alcohols (cf., however, Maruta and Iwama[(3)] who claim that ethylene glycol, glycerol, pentaerythritol, and sorbitol give a red colour reaction).

Buscarόns and Paraira[(6)] refined the reagent by preparing the piperidine salt of the vanadium oxine complex. In a beaker, stir 4 g of the complex with 50 ml piperidine for 3 h, then leave the beaker for 24 h in a hood. Separate the reddish-brown crystals by filtration, wash with ether, and dry in vacuum. Add 10 mg of the reagent to 0·5–1 ml of the alcohol (for solid alcohols use a benzene solution). After 1–2 min a wine-red colour appears which on addition of 1 drop of acetic acid turns deep red. Phenols and enols do not react.

(*d*) *4-Dimethylaminobenzaldehyde*

Colour reaction with 4-dimethylaminobenzaldehyde. Lipina[(7)] recommends the following procedure for the detection of traces of alcohols. Dissolve the sample in 75% sulphuric acid. To 3 ml of the solution add 0·2 ml of a freshly prepared 5% solution of 4-dimethylaminobenzaldehyde in the same solvent and heat the mixture for 15 min on a water-bath. Development of a colour (maximum absorption at 495 nm) indicates the presence of an alcohol. As esters are hydrolyzed in the procedure, the presence of an ester will invalidate the alcohol test. Sensitivity $1–7 \times 10^{-3}$ mg in 3 ml.

(*e*) *The Xanthate Reaction*

This reaction, originally suggested by Bamberger and Lodter,[(8)] is given by most primary and secondary alcohols and occasionally by tertiary alcohols.

A pellet of solid potassium hydroxide is added to a few drops of the substance and 0·5 ml of carbon disulphide in a test-tube. Warm the mixture on a water-bath at 60°C for 2 min (remember that carbon disulphide is highly inflammable). The potassium hydroxide usually dissolves and the solution assumes a yellow colour. Dilute the mixture with water, which produces a turbidity due to emulsified carbon disulphide. Despite the turbidity (which may separate out in droplets) add 1–2 drops of an aqueous solution of copper(II) sulphate and observe the precipitate. If an alkyl xanthate has been formed the precipitate is yellow or brown; if no xanthate is present it is blue (or dark, if copper(II) sulphide is formed in reaction between $copper^{2+}$ ions and carbon disulphide).

An excess of copper(II) sulphate must be avoided as the xanthate reaction may then be concealed by the dark precipitate of copper(II) sulphide.

The reactions are:

$$ROH + KOH + CS_2 \longrightarrow ROCSSK + H_2O$$

$$4ROCSSK + 2CuSO_4 \longrightarrow 2ROCSSCu + (ROCSS)_2 + 2K_2SO_4$$

The potassium alkyl xanthates may be used for paper or thin-layer chromatographic spot test detection and separation of alcohols.

For characterization of alcohols as 4-bromophenacyl xanthates and for estimation of alcohols acidimetrically or iodometrically, see Parts B and C, respectively.

## 2. Primary Alcohols

### (*a*) *Oxidation to Aldehydes*

Primary alcohols may be oxidized to the corresponding aldehydes or acids. With potassium permanganate in alkaline solution the alcohol is usually oxidized to the acid; with chromic acid it is often possible to isolate the aldehyde through its volatility. Good results are often obtained by using Beckmann's solution(9) for the oxidation.

This solution is prepared by dissolving 60 g potassium (or sodium) dichromate in 300 ml water and adding 50 g (27 ml) of concentrated sulphuric acid. Reflux for a short time 1 ml of the alcohol with 10 ml of Beckmann's solution, representing *ca.* 35 m-equiv. of oxygen or *ca.* 17 m-moles of alcohol oxidized to aldehyde. During reflux, pass any vapours or gases evolved through a solution of 2,4-dinitrophenylhydrazine or 4-carboxyphenylhydrazine in hydrochloric acid; after the reflux period distil the reaction mixture and collect a few ml of distillate in the solution used for trapping aldehyde vapours during reflux. The derivatives thus formed are useful for identification of the alcohol through m.p., paper- and thin-layer chromatography.

Both primary and secondary alcohols give carbonyl compounds on oxidation. If the distillate from the oxidation reduces Fehling's solution the carbonyl compound is an aliphatic aldehyde and the alcohol is primary. When the distillate reacts with 2,4-dinitrophenylhydrazine but not with Fehling's solution the oxidation product may be an aromatic aldehyde or a ketone, and the original hydroxy compound was either an aromatic primary alcohol or an aromatic or aliphatic secondary alcohol.

Another difference is that aldehydes, both aliphatic and aromatic, react with dimedone (5,5-dimethylcyclohexane-1,3-dione or 5,5-dimethyldihydroresorcinol).

Stensiö and Wachtmeister(10) claim that a chromium(VI) oxide–pyridine complex in glacial acetic acid solution is more selective than the Beckmann solution.

Prepare the reagent by adding chromium(VI) oxide (20 g, 0·2 mole) in portions with stirring to a solution of pyridine (31·6 g, 0·4 mole) in glacial acetic acid (150 ml), maintaining a temperature of 15–20°C. The reagent is made 0·8 M in chromium(VI) oxide by dilution with glacial acetic acid to 250 ml. It is sensitive to light and should be stored in darkness, preferably for not longer than 2 months.

Qualitative tests for alcohols with the reagent. Add 0·3 ml reagent to 20–30 mg or 0·03 ml sample. The presence of a primary or a secondary alcohol is revealed by a pronounced darkening of the mixture within 1–2 min (in some cases somewhat longer). Dilute the brown or brownish-violet solution with 2 ml water and extract once with 0·75 ml diethyl ether. Add 0·1 ml of the ethereal layer to 0·1 ml 0·25 M 2,4-dinitrophenylhydrazinium phosphate (dissolve 50 g 2,4-dinitrophenylhydrazine in 600 ml 85% phosphoric acid and dilute with ethanol to 1000 ml). A yellow or red precipitate within 5 min is regarded as a positive test.

The presence of an aldehyde is investigated, using Schiff's reagent, a solution of fuchsin, decolorized with sulphurous acid(11) or Fehling's solution (only aliphatic aldehydes).

This procedure is applicable only to pure or slightly humid alcohols, not to their aqueous solutions.

### (*b*) *Reaction with Nessler's Reagent*

Primary (and secondary) alcohols will reduce Nessler's reagent,(12) a mixture of equal volumes of 2 N sodium hydroxide and potassium mercury(II) iodide solution (dissolve 13·4 g mercury(II) chloride and 50 g potassium iodide in 1000 ml water). Add a few drops of the substance to 5 ml of this reagent and boil the mixture for a short time. The presence of a primary or secondary alcohol is revealed by the formation of a brownish-yellow to grey precipitate which turns grey (mercury) on standing.

### (*c*) *Colour Reaction with N-Bromosuccinimide*(13)

When added to a solution of N-bromosuccinimide, primary alcohols give a stable colour, secondary alcohols a fleeting colour, and tertiary alcohols no colour reaction.

## 3. Secondary Alcohols

### (*a*) *Oxidation to Ketones*

On oxidation with Beckmann's solution, secondary alcohols are converted to ketones. The method for oxidizing the alcohol and isolating the carbonyl compound is described above (p. 6). Critchfield and Hutchinson(14) describe conditions for the oxidation with acidic potassium dichromate, allowing secondary alcohols to be detected in the presence of primary alcohols, which are oxidized to acids.

(*b*) *Reaction with Nessler's Reagent*

Secondary alcohols react similarly to primary alcohols in reducing Nessler's reagent (see p. 7, para (*b*)).

## 4. Tertiary Alcohols

(*a*) *Oxidation*

Tertiary alcohols are oxidized only with difficulty. When oxidation does take place the carbon chain is ruptured and more than one oxidation product (ketones or acids) is formed; all of these have fewer carbon atoms than the original alcohol.

Bordwell and Wellman[15] make use of the resistance against oxidation to distinguish tertiary alcohols from primary and secondary.

When a drop of a solution of 25 g chromium(VI) oxide in 25 ml concentrated sulphuric acid, diluted with water to 100 ml, is added to a solution of a drop (or 10 mg) of the test substance in 1 ml acetone (preferably distilled over potassium permanganate) a greenish colour develops and an effervescence sets in within 2 sec when the substance is a primary or a secondary alcohol; the solution remains clear and without colour change for at least 1 min when the substance is a tertiary alcohol. An effervescence but no colour change may occur on prolonged reaction.

(*b*) *Reaction with Nessler's Reagent*

Tertiary alcohols do not reduce Nessler's reagent (see p. 7). A white precipitate may be formed on mixing the alcohol and the reagent, but it dissolves on shaking the test-tube. A dark precipitate may, however, be formed if the boiling is continued for some time.

(*c*) *Reaction with Denigès' Reagent*

Tertiary alcohols are readily dehydrated when boiled with Denigès' reagent;[16] alkenes, which give a yellow precipitate with the reagent, are formed.

The reagent is prepared by dissolving 50 g mercury(II) oxide in 200 g concentrated sulphuric acid and adding the solution to 1000 ml water. A few drops of the alcohol are heated with 2–3 ml of the reagent. A yellow precipitate is usually formed before the solution boils, and the test is only regarded as positive if the yellow precipitate has been formed within 1–2 min after starting the heating.

The reaction is thus a reaction of alkenes, not of tertiary alcohols. Primary alcohols are not dehydrated by the reagent, but some secondary alcohols are dehydrated almost as readily as the tertiary alcohols; such secondary alcohols will therefore give a positive reaction with Denigès' reagent. Other substances may also give a positive reaction if they are dehydrated by the reagent, e.g. ethers with at least one secondary or tertiary carbon atom at the oxygen bridge. Tertiary alcohols of the triphenylmethanol type, with no hydrogen atom available for the elimination of water, give a negative reaction.

The composition of the precipitate is probably $C_nH_{2n}\left(O\begin{smallmatrix}Hg\\ \\Hg\end{smallmatrix}SO_4\right)_3$.

The complex is not always stable and may be converted into a colourless precipitate, accompanied by gas evolution.

(*d*) *Reaction with Concentrated Hydrochloric Acid*

Tertiary alcohols are converted into *alkyl chlorides* by shaking 1 vol with 5–6 vol concentrated hydrochloric acid for 2 min. The alkyl chloride separates out to the surface of the originally homogeneous solution, but solutions containing primary or secondary alcohols usually remain homogeneous if the alcohol is soluble in hydrochloric acid. If, however, the alcohol is insoluble in hydrochloric acid no homogeneous phase is formed, even temporarily.

The alkyl chlorides are easily converted into S-alkylthiouronium picrates. See Par B, Chapter 6.2, p. 39 and recommended method *T*, p. 62.

## 5. Polyhydric Alcohols

Discussion in this paragraph will be restricted to the reactions of di- and trihydric alcohols as discussion of the reactions of polyols with four, five, six or more hydroxyl groups will be treated more naturally in the volume on carbohydrates in this series of Monographs in Organic Functional Group Analysis.

(*a*) *Formation of Boric Acid Complexes*

Polyhydric alcohols that have, on adjacent carbon atoms, at least two hydroxyl groups which are not forced to be *trans* to each other will, as stated by Böeseken and Hermans,(17) form complexes with boric acid of the structure

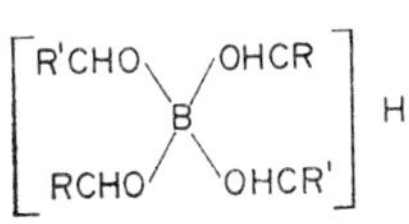

The complex acids are stronger than boric acid itself, and whereas an aqueous solution of sodium borate is strongly alkaline to phenolphthalein, the sodium salts of the complex acids are acidic to this indicator. It is therefore possible to test a substance for the presence of adjacent hydroxyl groups by adding to it a drop of phenolphthalein solution and just sufficient sodium hydroxide to make the solution alkaline (usually one drop of dilute sodium hydroxide is sufficient, but if acid groups are present these have to be neutralized before testing for adjacent hydroxyl groups). The weakly alkaline solution is then added to an aqueous solution of sodium borate, coloured red by phenolphthalein. If the substance is a 1,2-dihydroxy compound the two red solutions will turn colourless on mixing.

(*b*) *Reaction with Copper(II) Sulphate in Alkaline Solution*

Polyhydric alcohols give dark blue solutions containing complex copper(II) ions when treated with copper(II) salts and the solutions made strongly alkaline. An example is the well-known Fehling's solution.

The substance to be tested is dissolved in dilute sodium hydroxide and one or two drops of copper(II) sulphate solution are added. Copper(II) hydroxide may be precipitated, but redissolves on the tube being shaken if the substance is a polyhydric alcohol.

(*c*) *Perchromic Acid Test*

Substances of the polyethylene glycol type may be detected in the perchromic acid test as indicated by Springer and Isak.(18) On a spot plate 1 drop of the substance, 1 drop of 5% aqueous potassium dichromate and 1 drop of 4% sulphuric acid containing 3% hydrogen peroxide are mixed. On another location on the spot plate only potassium dichromate and the sulphuric acid solution are mixed.

The blue colour of perchromic acid will change to green in a few minutes in the blank test, whereas the colour will remain blue for 15–20 min in the presence of a polyethylene glycol. Solid polyethylene glycols are dissolved in glacial acetic acid prior to testing.

Further reactions of polyols will be mentioned in Parts B and C.

## 6. Other Aliphatic Hydroxy Compounds

If hydroxyl is not the only functional group present in the molecule it may be easier to base the identification of the compound on the other group, e.g. in many hydroxy acids, amino alcohols and hydroxy aldehydes or ketones. Reactions for such compounds should be sought in other volumes of the series of Monographs in Organic Functional Group Analysis. Many of the reactions mentioned above may be used, however, for the detection of the hydroxyl group in such compounds.

Substances containing a methylene group between a carbonyl group and another electronegative group show a tendency to tautomerize, exhibiting keto- and enol-forms. The hydroxy group of the enol usually gives a strong colour reaction with iron(III) chloride and, in distinction to the iron(III) chloride reaction of phenols (see Chapter 2, para 1(*a*)) the colour is more pronounced when the substance is dissolved in ethanol or other organic solvents rather than in aqueous solution. This is in accordance with the fact that weakly dissociating (i.e. weakly polar) solvents favour enolization.

## 7. Alicyclic Hydroxy Compounds

Alicyclic hydroxy compounds are secondary or tertiary alcohols; most of the reactions mentioned above for these types of hydroxy compounds are given also by the alicyclic hydroxy compounds.

CHAPTER 2

# AROMATICALLY BOUND HYDROXYL

## 1. Non-specific Reactions and Monohydric Phenols

### (*a*) *Colour Reaction with Iron*(*III*) *Chloride*

Most phenols give an intense colour reaction with iron(III) chloride; this is usually blue or violet, but, for some substituted phenols, it may be red or green. Empirical generalizations, having no absolute validity, are: (i) substituted phenols with two hydroxyl groups in the *ortho*-position give a green colour reaction; (ii) a carboxyl group *ortho* to a hydroxyl group gives a violet colour; (iii) a carboxyl group *para* to a hydroxyl group gives yellow or red, *o*-substitution increasing, *p*-substitution decreasing the intensity of the colour. *p*-Hydroxycarboxylic acids may also give yellow or reddish precipitates with iron(III) chloride. (iv) *meta*-Substituted phenols give either a weak colour reaction or none at all, but *m*-dihydroxybenzene gives an intense violet colour. (v) Many *para*-substituted phenols give normal reaction; cf., however, the above-mentioned reaction of *p*-hydroxycarboxylic acids and the oxidation of *p*-dihydroxybenzene to quinhydrone (see p. 17). (vi) Hydroxybenzenesulphonic acids often give a purple or red colour. (vii) The presence of a nitro group usually weakens the intensity of the reaction considerably, and nitrophenols often give no colour reaction at all.

The colour reaction with iron(III) chloride is considerably less sensitive in ethanolic than in aqueous solution (cf. the iron(III) chloride reaction of enols mentioned on p. 77). According to Soloway and Wilen(19) it may be made more sensitive by carrying it out in chloroform instead of in aqueous solution. One drop (*ca.* 30 mg) of the substance is dissolved in 1 ml chloroform, 2 drops of a solution containing 1 g anhydrous iron(III) chloride in 100 ml chloroform are added, followed by 2–3 drops of a solution containing 8 ml pyridine in 100 ml chloroform. The colour reaction is negative only for sterically hindered phenols and for some hydroxycarboxylic acids; even some nitrophenols give a positive colour reaction.

According to Pesez and Bartos[20] a mixture of potassium hexacyanoferrate(III) (potassium ferricyanide) and iron(III) chloride is more sensitive but less specific than the simple iron(III) chloride reaction. The colour is due to a reduction of iron(III) to iron(II) (or of hexacyanoferrate(III) ions to hexacyanoferrate(II) ions) and the reaction is therefore given by all compounds able to effect this reduction, e.g. aromatic amines and indoles.

The reagent is prepared (in darkness) by mixing equal volumes of 0·6% potassium hexacyanoferrate(III) and 0·9% iron(III) chloride hexahydrate. This reagent (0·2 ml) is added to an ethanolic solution of the sample (1 ml). After 5 min in darkness, 3 ml of 0·1 N hydrochloric acid is added and the mixture is left to stand for 15 min; a colour with absorption maximum at 720 nm develops. Sensitivity of the test: 3·5–13 $\gamma$ of a phenol in 1 ml may be detected.

### (*b*) *Liebermann's Reaction*[21]

Phenols having the *para*-position free form intensely coloured condensation products with concentrated sulphuric acid containing nitrous acid. The reaction is a nitrosation followed by condensation, and substances of the indophenol type are probably formed according to the scheme:

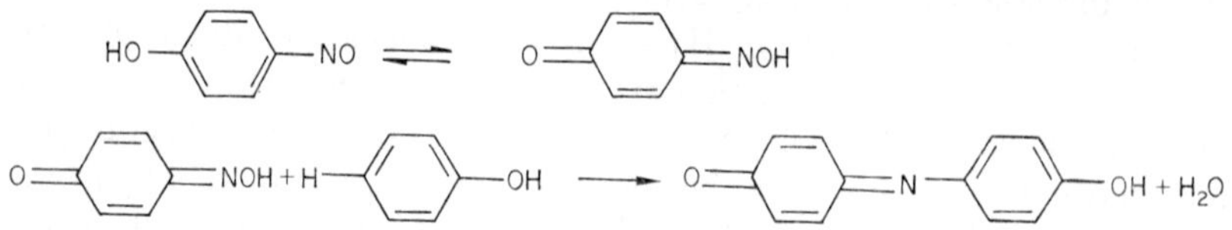


The condensation products resemble indophenols in being soluble in sodium hydroxide, forming intensely green or blue coloured solutions.

A few crystals (0·1–0·2 g) of sodium nitrite are dissolved in 1–2 ml concentrated sulphuric acid and 0·1–0·2 g of the substance is added. The mixture usually assumes an intense blue or green colour; if not, the mixture is heated for 1–2 min to a temperature not exceeding 100°C. When cool, the mixture is poured into about 20 ml water, giving a deep red solution, which on addition of excess sodium hydroxide turns blue or green. The reaction is positive only if both colour changes are seen; the indophenols formed act as acid–base indicators.

Turney[22] found that the test is subject to fairly severe limitations. Not only must the *p*-position be unoccupied, but the presence of different substituents may cause interference—groups which have an activating effect stronger than that of alkyl groups (e.g. hydroxyl-, amino- or methoxy groups with the exception of resorcinol and resorcinol monomethyl ether) and groups which have a deactivating effect stronger than that of the halogens (e.g. nitro-, aldehyde- or acetyl groups, with the exception of *o*-(chloromercury)phenol) give no positive response in the Liebermann reaction.

These limitations mean that usually only alkyl- or halogeno-substituted phenols with the *para*-position free will give a positive Liebermann reaction.

### (*c*) *Spot Test with Hexacyanoferrate(III)*

Inglett and Lodge[23] recommend the following procedure:

A drop of the sample solution is spotted onto a white porcelain spot plate and one drop of each of an ammonium chloride buffer (pH 10), 0·3% ethanolic 4-aminoantipyrin and 2% aqueous potassium hexacyanoferrate(III) (prepared freshly each time) are added successively. The coloured spot is compared with the yellow spot given by the reagent alone.

### (*d*) *Coupling with Diazotized Primary Amines*

Phenols with free *ortho*- or *para*-positions couple with diazotized primary aromatic amines when the free position is not *meta* to an amino- or to another hydroxy-group. Red or orange azo dyes, soluble in alkali, insoluble in acid, are formed in the coupling process. Sulphanilic acid or *p*-nitroaniline are generally used to form the diazo component. The diazotized amine is added to an alkaline solution of the phenol and an intensely coloured solution is immediately formed, from which the azo dye is precipitated on acidification (cf. the alcohol reaction with diazotized sulphanilic acid, p. 3). If amino groups *meta* to the hydroxy group are present in the phenol the azo dye may be soluble in acids, but it is still insoluble at neutral reaction.

Inglett and Lodge[23] have developed the reaction as a spot test.

### (*e*) *Reaction with Bromine Water*

Phenols with free *ortho*- or *para*-positions decolorize bromine water, giving substitution products which are usually precipitated. They may, after recrystallization (from ethanol) be used as derivatives for the characterization of individual phenols (see Part B, Chapter 6.5, p. 41).

### (*f*) *Solubility in Sodium Hydroxide*

Most phenols are soluble in dilute sodium hydroxide but exceptions, the so-called "crypto"-phenols, are known. In these the hydroxyl group is situated between highly branched side-chains or other substituents of strongly electron-donating character. Phenols are either insoluble or only moderately soluble in water; when insoluble in water they are also insoluble in a solution of sodium hydrogen carbonate unless they are substituted with electronegative substituents.

Alkaline solutions of phenols darken when exposed to the air, as a result of oxidation accompanied by polymerization.

Alkaline solutions of *ortho*- and *para*-nitrosubstituted phenols are strongly coloured, solutions of *o*-nitrophenols being orange, those of *p*-nitrophenols yellow.

## 2. Dihydric Phenols

(*a*) *ortho-Dihydroxy Compounds*

(i) *o*-Dihydroxybenzenes give an intense green colour with iron(III) chloride (cf. p. 13).

(ii) Alkaline solutions of *o*-dihydroxybenzenes are rapidly oxidized by air, turning dark brown in colour.

(iii) The *o*-dihydroxybenzenes react with acid chlorides, alkylenedihalides and diamines, forming heterocyclic compounds.

These reactions are not of great value for the detection of *ortho*-dihydroxybenzenes, but the reagents mentioned are used for the detection and characterization of other groups of compounds, and it must therefore be remembered that *ortho*-dihydroxybenzenes may also react with them.

(*b*) *meta-Dihydroxy Compounds*

(i) *m*-Dihydroxybenzenes give an intense violet colour with iron(III) chloride (cf. p. 13).

(ii) Alkaline solutions of *m*-dihydroxybenzenes are only slowly, and then slightly, coloured on air oxidation.

(iii) *m*-Dihydroxybenzenes, when heated with phthalic anhydride and a dehydrating agent (concentrated sulphuric acid, anhydrous zinc chloride) form phthaleins by condensation (Baeyer).[24] The phthaleins show a very strong fluorescence in alkaline solution. Substitution in the *meta*-position to the hydroxy groups may prevent the formation of a phthalein.

(iv) *m*-Dihydroxybenzenes may react with carbonyl group reagents, the tautomeric forms oxocyclohexadiene or dioxocyclohexene being present in solutions of the *m*-dihydroxybenzenes. On alkylation both *C*-alkylated derivatives of the oxo- or dioxo-form and *O*-alkylated derivatives of the dihydroxy-form may be produced.

(v) Among dihydroxybenzenes only the *meta*-compounds will couple regularly with diazotized amines to give azo dyes, whereas *ortho*- and particularly *para*-dihydroxybenzenes are oxidized frequently by nitrous acid, forming dark-coloured products (Griess,[25] Witt[26, 27]).

The diazo solution is added to a 1–2% aqueous solution of the compound, followed by sodium hydroxide until the solution is alkaline. If the substance is a *m*-dihydroxy compound, a clear solution of an azo dye will result; if it is an *o*- or a *p*-dihydroxybenzene the alkaline solution will usually be dark coloured, and a dark precipitate will soon be formed.

(*c*) *para-Dihydroxy Compounds*

(i) *p*-Dihydroxy compounds give an immediate blue colour reaction with iron(III) chloride. The blue colour rapidly changes to brownish-yellow because the iron(III) ions will oxidize the dihydroxybenzene to a quinone which forms a quinhydrone with the unoxidized compound. If more iron(III) chloride is added the quinhydrone may be precipitated as black-green crystals with metallic lustre.

(ii) With other oxidizing agents *p*-dihydroxybenzenes form quinones, often with the quinhydrones as isolable intermediates.

The quinones are often volatile in steam.

(iii) With hydroxylamine, *p*-dihydroxybenzenes may form mono- or dioximes of the corresponding quinones.

## 3. Trihydric Phenols

(*a*) *Vicinal Phenols*

(i) Vicinal phenols, dissolved in ethanol or water, give a reddish colour with a trace of a solution of iodine in ethanol. The colour may be so weak that it is necessary to compare the test solution with an equal volume of water to which the same amount of the iodine solution has been added.

(ii) Alkaline solutions of vicinal phenols are oxidized by air even more rapidly than *ortho*-dihydroxybenzenes. If an aqueous solution of the vicinal phenol is cautiously made alkaline the darkening of the solution may be seen to start at the surface of the liquid.

(iii) Alkylation yields a mixture of *O*- and *C*-alkylated products, showing the existence of tautomeric hydroxy- and oxo-forms.

(*b*) *Asymmetric Phenols*

The asymmetric phenols exhibit some of the character of hydroquinones, and several of the reactions mentioned for *p*-dihydroxybenzenes may also be used here. On alkylation, a mixture of *O*- and *C*-alkylated products is formed, showing the existence of tautomeric hydroxy- and oxo-forms.

(*c*) *Symmetric Phenols*

(i) Symmetric phenols, in concentrated hydrochloric acid, give a red to purple colour reaction with a pine splint, provided there is still an unsubstituted hydrogen atom in the molecule.

(ii) In symmetric phenols the tautomeric cyclohexanetrione form is predominant, and on alkylation in alkaline solution *C*-alkylated derivatives are formed almost exclusively. The methoxy compounds may be prepared by methylation with diazomethane.

(iii) Symmetric phenols, according to Benedict,[28] form quinone oximes with nitrous acid.

CHAPTER 3

# HYDROXYL IN HETEROCYCLIC COMPOUNDS

## 1. Five-membered Systems

(*a*) *Aromatic Systems*

Hydroxy compounds of five-membered heterocyclic systems are very unstable, being rapidly polymerized to dark, high molecular weight compounds. Very few remarks of interest for the detection and estimation of these compounds can be made.

(i) *Furan.* According to Elderfield and Dodd,[(29)] hydroxyfurans are extremely unstable if the furan nucleus is not electronegatively substituted in both 2- and 5-positions. According to Katritzky and Lagowski[(30)] 2-hydroxyfuran is unknown, even when the furan nucleus is substituted in other positions. 3-Hydroxyfuran does exist but is tautomeric with 3-oxo-2,3-dihydrofuran. The hydroxy form gives a colour reaction with iron(III) chloride and adds on bromine.

Furans with hydroxy groups in a side-chain will often behave as aliphatic hydroxy compounds.

(ii) *Thiophen.* According to Blicke,[(31)] both 2- and 3-hydroxythiophen are unstable, and according to Hartough[(32)] they exist almost exclusively in the oxo-form, e.g.

5-methyl-2-hydroxythiophen $\rightleftharpoons$ 5-methyl-4-thiolen-2-one.

Compounds with hydroxy groups in side-chains are similar to the analogues in benzene chemistry with the exception of 2-hydroxymethylthiophen, which polymerizes to a thiophen-formaldehyde resin.

(iii) *Pyrrole.* All hydroxypyrroles are unstable.

(iv) 5-*Membered systems with* 2 *or more heteroatoms.* Here the pyrazole and imadazole nuclei have been studied most thoroughly. Many hydroxy derivatives of these compounds give a colour reaction with iron(III) chloride, but on alkylation mixtures of *O*-, *C*- and *N*-alkyl derivatives are formed, showing the existence of tautomeric forms.

(*b*) *Non-aromatic Systems*

Hydroxy compounds of this type usually behave similarly to alicyclic hydroxy compounds. When the hydroxy group is on a carbon atom with a double bond to a heteroatom or to another carbon atom the corresponding oxo-form is present, and is often the main component in the equilibrium mixture.

## 2. Six-membered Systems

(*a*) *Aromatic Systems*

(i) *Hydroxypyridines and hydroxyquinolines* give a red, brownish-red, blue, or green colour reaction with iron(III) chloride.

(ii) Of the hydroxypyridines, and hydroxyquinolines with the hydroxy group in the pyridine nucleus, the 3-hydroxy compounds behave as phenols, coupling with diazotized sulphanilic acid to form azo dyes.

(iii) 2- and 4-Hydroxypyridines and quinolines are tautomeric with the corresponding oxo-forms of the dihydro-systems. On *alkylation* a mixture of *O*- and *N*-alkylated products is formed.

(*b*) *Non-aromatic Systems*

Here, as in the 5-membered rings, the hydroxy compounds behave similarly to the alicyclic hydroxy compounds. In the partially hydrogenated systems tautomerism with oxo compounds of further hydrogenated systems has to be considered.

## 3. Seven- and Higher-membered Systems

More or less hydrogenated aza-, oxa-, sulpha- and diaza-epines are becoming more and more important. The properties of hydroxy compounds of such systems are similar to those of other types of heterocyclic hydroxy compound. Details should be ascertained from monographs dealing with the particular heterocyclic system involved.

# REFERENCES TO PART A


1. Meyer, H., *Analyse und Konstitutionsermittelung*, 1st Edn.
2. Blair, A. T. and Pantony, D. A., *Anal. Chim. Acta* **13**, 1 (1955).
3. Maruta, S. and Iwama, F., *Nippon Kagaku Zasshi* **80**, 1131 (1959); *Chem. Abs.* **55**, 4254 i (1961).
4. Buscaróns, F., Marin, J. L. and Claver, J. J., *Anales real. soc. españ. fis. y quim.* **49B**, 367 (1953); *Chem. Abs.* **49**, 80 c (1954).
5. Kudo, I. and Aoki, I., *Bunseki Kagaku* **6**, 791 (1957); *Chem. Abs.* **53**, 4013 h (1959).
6. Buscaróns, F. and Paraira, M., *Chim. Anal.* (*Paris*) **49**, 401 (1967).
7. Lipina, T. G., *Novoe v Oblasti Sanit-Khim. Analiza* (*Raboty po Prom. Sanit. Khim.*) **1962**, 150; *Chem. Abs.* **59**, 4471 e (1963).
8. Bamberger, E. and Lodter, W., *Ber.* **23**, 213 (1890).
9. Beckmann, E., *Liebigs Annalen* **250**, 325 (1889).
10. Stensiö, K. E. and Wachtmeister, C. A., *Acta Chem. Scand.* **18**, 1013 (1964).
11. Schiff, H., *Compt. rend. Acad. Sci.* **61**, 45 (1865); **64**, 182 (1867); *Liebigs Annalen* **140**, 92 (1866).
12. Rosenthaler, L., *Arch. Pharm.* **244**, 373 (1906); *Angew. Chem.* **20**, 412 (1907).
13. Severin, M., *Bull. Rech. Agron. Gembloux* **1**, 289 (1966); *Chem. Abs.* **67**, Ref. 17618 (1967).
14. Critchfield, F. E. and Hutchinson, J. A., *Anal. Chem.* **32**, 862 (1960).
15. Bordwell, F. G. and Wellman, K. M., *J. Chem. Educ.* **39**, 308 (1962).
16. Denigès, G., *Compt. rend. Acad. Sci.* **126**, 1145; 1277 (1898).
17. Böeseken, J. and Hermans, P. H., *Rec. Trav. chim. Pays-Bas* **40**, 525 (1921).
18. Springer, R. and Isak, H., *Z. anal. Chem.* **199**, 363 (1963).
19. Soloway, S. and Wilen, S. H., *anal. Chem.* **24**, 979 (1952).
20. Pesez, M. and Bartos, J., *Ann. Pharm. France* **23**, 281 (1965).
21. Liebermann, C., *Ber.* **7**, 247; 806; 1098 (1874).
22. Turney, T. A., *J. Org. Chem.* **22**, 1692 (1957).
23. Inglett, G. E. and Lodge, J. P., *Anal. Chem.* **31**, 248 (1959).
24. von Baeyer, A., *Liebigs Annalen* **183**, 1 (1876).
25. Griess, P., *Ber.* **15**, 2190 (1882); **17**, 338 (1884).
26. Witt, O. N. and Mayer, F., *Ber.* **26**, 1072 (1893).
27. Witt, O. N. and Johnson, E. S., *Ber.* **26**, 1908 (1893).
28. Benedict, R., *Ber.* **11**, 1375 (1878).
29. Elderfield, R. C. and Dodd, Jr., T. N. in Elderfield, *Heterocyclic Compounds*, Vol. **I**, p. 177, Wiley (1950).
30. Katritzky, A. R. and Lagowski, J. M., *Advances in Heterocyclic Chemistry* **2**, 5 Academic Press (1963).
31. Blicke, F. F. in Elderfield, *Heterocyclic Compounds*, Vol. **I**, p. 229, Wiley (1950).
32. Hartough, H. D. in Weissberger, *The Chemistry of Heterocyclic Compounds*, Vol. **3**, p. 287, Interscience (1952).

# Part B

## Characterization of Hydroxy Compounds

In many instances it is possible to make a direct determination of hydroxy compounds, i.e. the determination of the equivalent weight per hydroxyl group or, more useful, e.g. for polymers, the percentage of hydroxyl in the compound. Frequently it is less complicated, however, to use a derivative, prepared for the characterization of the hydroxy compound, for the estimation of the equivalent weight of the compound. In this section, therefore, methods for the preparation of derivatives of hydroxy compounds will be described, together with methods for the chromatographic characterization of hydroxy compounds, often through chromatography of the derivatives mentioned.

CHAPTER 4

# ARYLURETHANES

HYDROXY compounds will react with isocyanates according to the scheme:

$$R{-}N{=}C{=}O + R'OH \rightarrow R{-}NH{-}COOR'$$

the reaction thus being an addition of the hydroxy compound at the —N=C= double bond. Thiols react analogously, forming thiourethanes. Also primary and secondary amines react with isocyanates, forming 1,3-disubstituted ureas.

## 1. Phenylurethanes

Goldschmidt and Meissler[(1)] recommend that the hydroxy compound be added to a 10% solution of phenylisocyanate in toluene and then to reflux the solution for 1–3 h. The solvent is removed by distillation and the residue crystallized from ligroin (b.p. 60–90°C) or petroleum ether (b.p. 40–60°C).

## 2. α-Naphthylurethanes

Willstätter and Hocheder[(2)] recommend the use of α-naphthylisocyanate instead of phenylisocyanate as the α-naphthylurethanes are less soluble and have higher m.p.'s than the phenylurethanes. The procedure for the preparation of these derivatives is analogous to the preparation of phenylurethanes.

According to Johnson[(3)] still less soluble and more readily crystallizable derivatives are obtained by using 2,4,6-triidophenylisocyanate as reagent.

## 3. *m*-Nitrophenylurethanes

The arylisocyanates are very sensitive to the humidity of the air; they react with water more readily than with most organic hydroxy compounds. Hoeke[(4)] and later Veibel *et al.*[(5, 6)] tried to avoid this complication by using, instead of the arylisocyanates, the corresponding carboxazides, which are not sensible to humidity. When heated to about 100°C, the carboxazides undergo a Curtius rearrangement:

$$\text{Ar—CO—N═}\overset{+}{\text{N}}\text{═}\overset{-}{\text{N}} \rightarrow \text{Ar—N═C═O} + \text{N}_2,$$

thus forming an isocyanate in the solution which may contain the substance under examination. *m*-Nitrobenzazide was the preferred reagent as it is readily prepared from ethyl benzoate through ethyl *m*-nitrobenzoate, and the resulting derivatives may be used for the titanometric estimation of the equivalent weight.

Mix 1–2 g of *m*-nitrobenzazide with an equivalent weight of the hydroxy compound in a 50-ml flask, add 10 ml xylene and reflux the mixture for 1–3 h. If the hydroxy compound is a tertiary alcohol, use toluene or ligroin (b.p. 60–100°C) as solvent instead of xylene in order to prevent dehydration of the alcohol, and extend the reflux time to 4–6 h. The derivatives often precipitate out on cooling and are then isolated by filtration. If no precipitation occurs they are isolated and purified as indicated for the phenylurethanes.

When the hydroxy compound and the solvent are not absolutely dry the *m*-nitrophenylurethane will be contaminated with bis(*m*-nitrophenyl)urea, a yellow, crystalline substance of m.p. 245–247°C which is formed, instead of *m*-nitrophenylurethane, from tertiary alcohols if the boiling point of the solvent is sufficiently high for the alcohol to be dehydrated.

Enols react readily with *m*-nitrobenzazide, forming enolic *m*-nitrophenylcarbamic esters.

As indicated by Veibel *et al.*[6] and Rekker *et al.*,[7] secondary reactions may take place with α- and β-hydroxy acids or esters of such acids, resulting in the formation of heterocyclic compounds (oxazolidinediones, tetra- or dihydrooxazinediones:

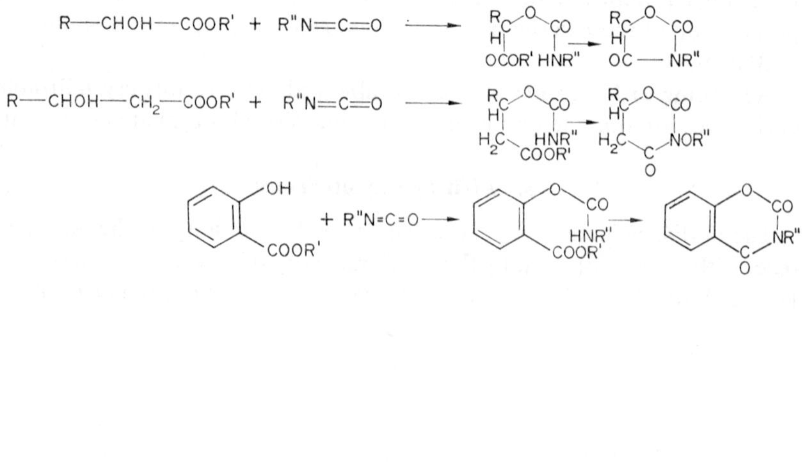

CHAPTER 5

# ESTERS

## 1. Acetates

As the acetates of lower alcohols and some phenols are liquid at room temperature, they are only useful for characterization by gas chromatography. This method, however, is becoming more and more important, and therefore methods for the preparation of acetates will be considered briefly. Acetic anhydride or acetyl chloride are the most prominent acetylating agents.

(*a*) *Acetic Anhydride.* Mix 0·5 g of the hydroxy compound with 1 ml acetic anhydride and add one drop of concentrated sulphuric acid. Boil the mixture for 1–2 min, then pour it into water. The acetate separates out when the excess acetic anhydride has been hydrolyzed, and may often be induced to crystallize by scratching the side of the test-tube.

Some acetates are more soluble in acetic acid, even when dilute, than in water. In order to isolate such acetates the solution is neutralized with sodium hydroxide and the neutral solution extracted with ether. The acetate is then isolated by evaporation of the ether from the ether layer.

Fritz and Schenk[(8)] recommend ethyl acetate as solvent and perchloric acid as catalyst for the acetylation. As this modification is suitable for quantitative estimation of hydroxy compounds it will be described in detail in Part C, Chapter 10.1, p. 77, and Method *E*, (*a*), p. 117.

(*b*) When *acetyl chloride* is used for the acetylation dissolve 0·5 g of the substance in 2 ml pyridine and add 2–2·5 ml acetyl chloride. Leave the mixture for 3–4 h at room temperature; then add 20 ml dilute sulphuric acid to neutralize the excess pyridine. Isolate the acetate as above.

Acetates of tertiary alcohols are difficult to prepare by this method, as the alcohols are dehydrated by the reagent or converted into alkyl chloride by the hydrogen chloride produced. By carrying out the acetylation with acetyl chloride in the presence of metallic magnesium, however, Spassow[(9)] obtained acetates of tertiary alcohols in good yields.

Dilute 1–2 g of the alcohol with 5–10 ml anhydrous ether and add 1 g magnesium powder. Dilute 2–3 ml acetyl chloride with 5 ml anhydrous ether and add this solution dropwise to the sample solution. The reaction may be rather vigorous and the test-tube is cooled in water to avoid too violent boiling. When all the acetyl chloride has been added, leave the test-tube for 1 h at room temperature, then reflux for 1–2 h. The reaction product is usually a white precipitate or a semi-solid mass. After dilution with ether it is poured into a mixture of ice and sodium carbonate solution. Separate the ether layer, and extract the aqueous layer with ether. Dry the combined ether layers over a mixture of anhydrous sodium sulphate and sodium carbonate and isolate the acetate by fractional distillation.

For the acetylation of phenols, too, the addition of magnesium powder may be useful. Dissolve 1–2 g of the phenol in 20 ml dry benzene, add 0·5–1 g magnesium powder and 1–2 ml acetyl chloride and heat the mixture at about 90°C for 0·5–1 h. Cool, dilute the mixture with ether, and isolate the acetate as above. In most cases it will crystallize when the solvents have evaporated, making a fractional distillation unnecessary.

## 2. Benzoates

Benzoates usually have higher m.p.'s than the corresponding acetates. The benzoates of all phenols and of many aliphatic hydroxy compounds are solid at room temperature. Benzoates are therefore of greater value than acetates for the characterization of hydroxy compounds.

A convenient method for preparing the benzoates of phenols is as follows:

Dissolve 0·5 g of the phenol in 3 ml anhydrous pyridine and add 0·5 g benzoyl chloride. A spontaneous reaction frequently starts and is maintained by refluxing the mixture for a few min. Cool, and pour the reaction mixture into 10 ml water with vigorous stirring, so that the benzoate precipitates. Decant the supernatant solution, and treat the residue with 5 ml sodium carbonate solution to remove benzoic acid, if present. Filter off the precipitate, wash with water, and recrystallize from ethanol or aqueous ethanol.

Most phenols may be benzoylated by the Schotten–Baumann method.[10,11] To 3–5 g benzoyl chloride add 1 g phenol, followed by excess 2 N sodium hydroxide, and shake the mixture well in a stoppered test-tube until the smell of benzoyl chloride has disappeared; loosen the stopper from time to time as the reaction is accompanied by evolution of heat. The reaction time is usually 5–10 min. The benzoate often separates as semi-solid lumps containing some unreacted benzoyl chloride. Free the lumps from benzoyl chloride and benzoic acid by grinding them with sodium hydroxide, and recrystallize the crude benzoate from ethanol or aqueous ethanol.

If the substance is a phenolcarboxylic acid the benzoate will be soluble in sodium hydroxide and no ester will appear to be formed. The benzoate, mixed with benzoic acid, will, however, be precipitated on acidification of the solution. Great care must be taken to separate the benzoate from benzoic acid, either by a chromatographic method, by passing the solution through an ion exchange resin, or by recrystallization.

For substances which are sensitive to alkali, either sodium carbonate or sodium hydrogen carbonate may be used instead of sodium hydroxide; alternatively, acid benzoylation may be employed. Heat the substance for some time with benzoyl chloride at a temperature in the neighbourhood of the b.p. of benzoyl chloride (198°C). The Schotten–Baumann method is, however, preferable if the substance is stable to alkali.

Electrophilic substituents make the benzoylation of hydroxy compounds more difficult. Mononitrophenols may be benzoylated with some difficulty by the Schotten–Baumann method; polynitrophenols are unaffected, however.

### 3. *p*-Nitrobenzoates

Esters of *p*-nitrobenzoic acid have higher m.p.'s than esters of benzoic acid, and even the esters of lower aliphatic alcohols are solid at room temperature. These esters are therefore preferable to benzoates as derivatives of the lower alcohols. Another advantage of *p*-nitrobenzoates is that they may be titrated with titanium(III) chloride and can thus be used for the quantitative estimation of the equivalent weight of the hydroxy compound (see Part C, Chapter 10.8 (*d*), p. 92 and Method *K*, p. 133).

Henstock[12] recommends the following procedure for the preparation of *p*-nitrobenzoates:

Dissolve 0·5 g of the alcohol in 5 ml water, or dilute the aliquot of the solution of the alcohol that corresponds to 0·5 g to a total volume of 5 ml. Add to the solution 0·5 g *p*-nitrobenzoyl chloride, 3 ml ethanol-free ether, and a trace of phenolphthalein (not dissolved in ethanol). Shake the mixture well and add dilute sodium hydroxide dropwise until the aqueous layer assumes a pink colour. Continue to shake for some min; then separate the ether layer and dry it over calcium chloride. Isolate the *p*-nitrobenzoate by evaporation of the ether and recrystallize from petroleum ether.

Komárek *et al.*[13] prepare *p*-nitrobenzoates by heating the alcohol with *p*-nitrobenzoyl chloride in sealed ampoules at 100°C for 30–60 min. This method is particularly useful for semimicro and micro quantities of hydroxy compounds.

For anhydrous compounds the pyridine method recommended above (p. 28) for the preparation of phenyl benzoates can be applied to the preparation of *p*-nitrobenzoates if *p*-nitrobenzoyl chloride is substituted for benzoyl chloride.

### 4. 3,5-Dinitrobenzoates

Esterification with 3,5-dinitrobenzoic acid was originally suggested by Reichstein,[14] and improved methods for the preparation of 3,5-dinitrobenzoates have been given by Saunders *et al.*,[15] by Dunbar and Ferrin[16] and by Huth[17] (see below, recommended Methods *A*, *B* and *C*). The 3,5-dinitrobenzoates have proved extremely useful for the characterization of hydroxy compounds by different chromatographic methods (paper, thin-layer and gas chromatography) (see below, Chapter 7.2 (iv) and Chapter 7.3 (iv), p. 48 and p. 54, respectively).

Garska *et al.*[18] recommend X-ray diffraction patterns for the identification of 3,5-dinitrobenzoates; m.p.'s and X-ray diffraction patterns are given for 24 esters of $C_1$–$C_{22}$ alcohols.

For column chromatographic separation and estimation of alcohols by ultraviolet spectroscopy, Jensen *et al.*,[19] van Duin[20] and Freytag and Baustian[21] recommend 4-dialkylamino-3,5-dinitrobenzoates, the advantage being that the derivatives are coloured and therefore visible without spraying. They are, according to van Duin,[20] prepared quantitatively as indicated in recommended Method *D*, p. 58 (Chapter 8.2).

### 5. (Substituted) 4-Phenylazobenzoates

Other coloured derivatives suggested for chromatographic separation and spectroscopic estimation of hydroxy compounds are 4-phenylazobenzoates and 4′-nitro-4-phenylazobenzoates. Pohloudek-Fabini *et al.*[22–24] recommend 4-phenylazobenzoates for the determination of free and (after saponification) esterified alcohols in essential oils by paper chromatographic separation and spectroscopic estimation of the derivatives, prepared as indicated in recommended Method *E*, p. 58.

4′-Nitroazobenzene-4-carboxylates were introduced by Hecker[25] and were used by Bosvik *et al.*[26] for the identification of alcohols by infrared spectroscopy and by X-ray diffraction patterns. For their preparation see recommended Method *F*, p. 58.

Amin and Samad[27] have studied the characterization of mono- and polyhydric phenols as 4′-nitrophenylazobenzoates (see recommended Method *G*, p. 58).

## 6. Acid Esters of Dibasic Acids

(*a*) *Alkyl Hydrogen Succinates*

Veibel and Pedersen[28] as well as Narang and Mathur[29] found that succinic anhydride is converted readily into alkyl potassium succinate when one equivalent of potassium or concentrated potassium hydroxide is added to a solution of the anhydride in the alcohol studied. The potassium salt is precipitated by addition of acetone and isolated by filtration. From its solution in water the acid ester is liberated by acidification with hydrochloric acid. Both the potassium salts and the acid esters can be used for the acidimetric or alkalimetric determination of the equivalent weight of the alcohol.

(*b*) *Alkyl Hydrogen Phthalates or Tetrachlorophthalates*

Veibel and Pedersen[28] also prepared acid esters of phthalic acid by the method indicated for preparation of alkyl hydrogen succinates. Both primary and secondary alcohols can be esterified by this method, but it fails for the preparation of acid phthalates of tertiary alcohols, just as the preparation of acetates of tertiary alcohols presents difficulties. Fessler and Shriner,[30] Lawlor[31] and Elving and Warshowsky[32] have shown that, when the tertiary alcohol is converted into a strongly nucleophilic reagent by reaction with ethyl magnesium bromide, the preparation of *tert.*-alkyl hydrogen phthalate (or tetrachlorophthalate) is possible. The following reaction scheme is suggested:

$$R_3COH + C_2H_5MgBr \longrightarrow R_3CO^- \overset{+}{Mg}Br + C_2H_6$$

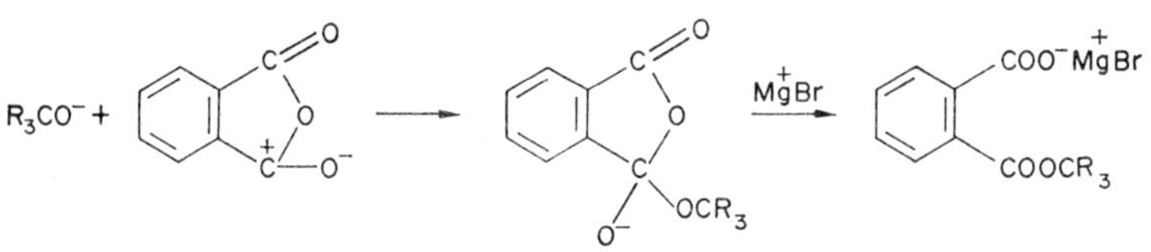

For the procedure see recommended Method *H*, p. 59.

3-Nitrophthalic anhydride has also been suggested as a reagent for the formation of acid esters, e.g. by Dickinson *et al.*,[33] by de Graef and Pierret[34] and by Floria *et al.*[35]

(*c*) *Dialkyl Dihydrogen Pyromellitates*

Siggia *et al.*[36–38] found that the dianhydride of pyromellitic acid (benzene-1,2,4,5-tetracarboxylic acid dianhydride) reacts faster than (substituted) phthalic anhydride, in fact as fast as acetic anhydride catalyzed by perchloric acid. Originally[36, 37] tetrahydrofuran was used as solvent, but later it was found that dimethylsulphoxide is a better solvent.[38] The reaction is:

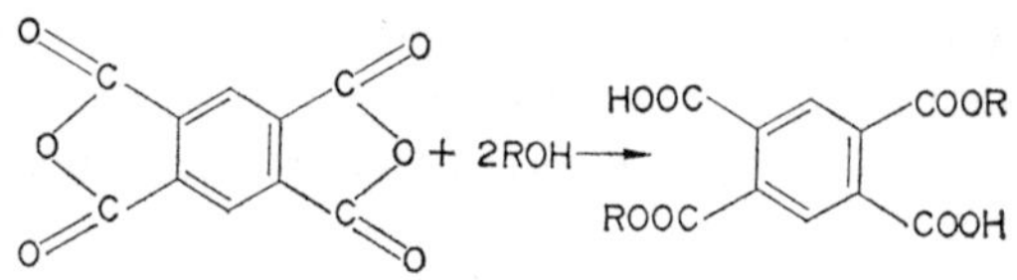


It is possible to isolate the acid esters, but usually the method is used for the estimation of hydroxyl groups, and the procedure recommended is therefore given in Part C, Chapters 10.1 (*f*), p. 80 and 11.2 (*i*), p. 122.

## 7. 3-Cyano-3,3-diphenylpropionates

Salmon-Legagneur and Neveu[39] recommend the preparation of these esters for the characterization of alcohols through m.p.'s and infrared spectroscopy, the absorption of the nitrile group being used for characterizing the esters. The reagent is 3-cyano-3,3-diphenylpropionyl chloride, prepared from the commercially available diphenylacetonitrile by a procedure described by Salmon-Legagneur.[40] For details see recommended Method *I*, p. 59.

## 8. Alkyl Methyl Esters of Oxalic Acid

Oppenauer[41] indicates that esters of all types of alcohols are formed when the alcohol is treated with the anhydride of methyl hydrogen oxalate in the presence of a very small excess of pyridine. The reagent,

$$H_3COOC—CO—O—CO—COOCH_3,$$

is prepared from dimethyl oxalate *via* methyl potassium oxalate by reaction with the acid chloride of methyl hydrogen oxalate. The procedure for the preparation of the reagent and its use for the estimation of alcohols is described in Part C, Chapter 10.1 (*h*), p. 80.

### 9. Anthranilates

Staiger and Miller[42] have described the reaction of isatoic anhydride with alcohols and phenols, leading to anthranilates with evolution of carbon dioxide. The reaction is a nucleophilic attack on the anhydride catalyzed by sodium hydroxide:

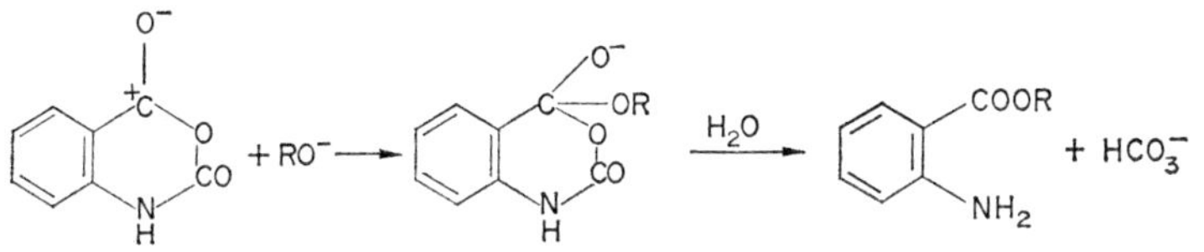


The anthranilates, which are often oily, may be characterized through their picrates. For details of preparation, see recommended Method *J*, p. 60.

Teeter[43] prepares *N*-methylanthranilates for mass-spectrometric characterization of alcohols. Amines, thiols, and compounds with reactive methylene groups interfere, as they, too, will react with the reagent producing *N*-methylanthranilates.

### 10. *p*-Toluenesulphonates

*p*-Toluenesulphonyl chloride is recommended by Mesnard *et al.*[44, 45] for the estimation of primary and secondary alcohols and of phenols; tertiary alcohols do not react with this reagent. The procedure to be followed is given in Part C, Chapter 10.1 (*i*), p. 81.

### 11. 2,4-Dinitrobenzenesulphenates

Kharasch *et al.*[46] recommend the characterization of all types of alcohols as esters of 2,4-dinitrobenzenesulphenic acid through their m.p.'s. Lefèbvre *et al.*[47] report on the characterization of these esters through thin-layer chromatography.

The reagent, 2,4-dinitrobenzenesulphenyl chloride, m.p. 95–96°C, is prepared by catalytic chlorolysis of the corresponding disulphide:

$$(NO_2)_2C_6H_3S-SC_6H_3(NO_2)_2+Cl_2 \rightarrow 2(NO_2)_2C_6H_3SCl$$

Dissolve in a dry 50-ml flask 1 g of 2,4-dinitrobenzenesulphenyl chloride, shaking it with 10 ml of absolutely dry ethylene dichloride. Add 1 g of the alcohol, shake the mixture to ensure complete dissolution and then add 1 ml anhydrous pyridine, causing a darkening of the yellow-coloured solution. After further shaking leave the mixture to stand at room temp for 15–30 min. Usually a white precipitate is formed. Filter it off and wash it with small quantities of hot ethylene dichloride. Combine the filtrate and washings and evaporate to dryness *in vacuo* at room temp. Extract the residue so obtained with 3–5 10-ml portions of cold water, filter the 2,4-dinitrobenzenesulphenate off at the pump, dry it in a vacuum desiccator (or a vacuum drying apparatus at 56°C) over calcium sulphate and recrystallize from absolute ethanol or an ethanol–ligroin mixture.

### 12. 2-Alkoxycarbonylbenzenesulphonic Acids

Iyer and Mathur[48] and Narang *et al.*[49] recommend 2-sulphobenzoic anhydride for the determination and microdetermination of primary and secondary $C_1$–$C_{18}$ alcohols, glycols, sugars, hydroxy acids, and free alcohol groups in polyglycol ethers. Amino compounds, too, react with the reagent. The procedure to follow is given in Part C, Chapter 10.1 (*j*), p. 82.

### 13. *p*-Bromophenacyl Alkyl Xanthates

Berger and Uldall[50] recommend the use of the potassium alkyl xanthates mentioned above (p. 5) for the preparation of *p*-bromophenacyl esters of the xanthic acids:

$$ROCSSK+BrC_6H_4COCH_2Br \rightarrow ROCSSCH_2COC_6H_4Br+KBr$$

Dissolve 0·6–1 g (4 m-moles) of potassium alkyl xanthate in 10 ml acetone and 1 ml water without heating; if the xanthate does not dissolve completely add a further 1–2 ml water. Add a solution of 1 g (3·6 m-moles) of *p*-bromophenacyl bromide in 10 ml acetone and leave the mixture at room temp for 5 min, then add 25–30 ml water, shake the mixture frequently during 5 min to dissolve potassium bromide, and filter off the *p*-bromophenacyl alkyl xanthate by suction. The ester is usually reasonably pure but may be recrystallized from ethanol.

If precipitation does not occur, cooling the solution and scratching the walls of the flask may induce crystallization.

The preparation of these esters is adaptable to both the micro or semimicro scale, down to 5 mg potassium alkyl xanthate.

The m.p. of 30 esters are recorded by Berger and Uldall.[50]

The m.p. of esters of xanthates of isomeric alcohols may sometimes coincide. In such cases the esters of *p*-phenylphenacyl alcohol or of *p*-chlorophenacyl alcohol may make a distinction possible. They are prepared analogously, using the corresponding bromides as the reagents.

### 14. Esters of Pyruvic Acid 2,6-Dinitrophenylhydrazone

Schwartz and Brewington[51] have prepared such esters of primary alcohols $C_1$–$C_{21}$, secondary alcohols $C_3$–$C_{19}$ (only alkan-2-ols) and tertiary alcohols $C_4$–$C_9$. The reagent 2,6-dinitrophenylhydrazine is prepared from chloro-2,6-dinitrobenzene and hydrazine. For details of procedure, see recommended Method *K*, p. 60.

### 15. Alkyl Nitrates

A method for the isolation of mono- and dihydric alcohols from complex mixtures involves conversion[52] of the alcohols into alkyl nitrates by treating the mixture with acetyl nitrate. The alkyl nitrates are more easily separated chromatographically from polar substances than are the alcohols as they are only slightly more polar than hydrocarbons. Using a column of silica gel, gram-quantities of the nitrates may be isolated. The nitrates may then be converted back into the alcohols by hydrogenation, e.g. with lithium aluminium hydride, and the alcohols can then be identified by one of the current methods.

### 16. Alkyl Hydrogen Sulphates

Borecký[53, 54] recommend preparation of the alkyl hydrogen sulphates for the paper chromatographic separation and identification of aliphatic $C_{10}$–$C_{18}$ alcohols.

Dissolve 1–2 drops of chlorosulphonic acid in 0·5 ml ether, and add one drop or the corresponding quantity of a solid alcohol dissolved in 0·5 ml ether. After 15 min at room temp add 1–2 drops of a saturated aqueous solution of sodium sulphate. After thorough mixing, apply the ethereal layer directly to one of the chromatographic methods mentioned below.

An excellent survey of acylation methods for hydroxy compounds is given by Mathur.[55]

CHAPTER 6

# OTHER DERIVATIVES

## 1. Ethers

### (*a*) *Methyl Ethers*

Methyl ethers are more important for the characterization of phenols than for the characterization of aliphatic and alicyclic hydroxy compounds as they are more easily prepared by methylation with dimethyl sulphate, methyl iodide, or diazomethane; these reagents will not convert alcohols into ethers.

Dissolve the phenol in approximately the equivalent quantity of sodium hydroxide; the use of N up to 6–8 N is satisfactory. Add in small portions slightly more than 1 mole dimethyl sulphate per equivalent of phenol. The methylation may give rise to so great an evolution of heat that cooling may be necessary. The reaction is usually complete in 15–30 min; excess dimethyl sulphate is then destroyed by addition of more sodium hydroxide. Dilute the reaction mixture with water, and isolate the methylated phenol by filtration or, if liquid, by separation in a tap funnel. Extract the aqueous layer with ether, combine the ether layer with the organic layer first separated, dry the ethereal solution with magnesium sulphate, remove the ether by evaporation and purify the aryl methyl ether by recrystallization, distillation or, if necessary, steam distillation.

When working with dimethyl sulphate it should be remembered that this substance is a dangerous poison, both as a vapour and in contact with the skin, through which it is readily absorbed. It is, however, easily and rapidly destroyed by dilute ammonia.

The methylation of polyhydric phenols and of heterocyclic hydroxy compounds may, as mentioned above (pp. 16, 17, 18, 19, 20), lead to the formation of a mixture of the methoxy compound and *C*- or *N*-methylated derivatives of the tautomeric oxo-forms.

The aryl methyl ethers may be used for the gas-chromatographic characterization of phenols.

### (*b*) *2,4-Dinitrophenyl Ethers*

For the characterization of phenols through m.p.'s of derivatives 2,4-

dinitrophenyl ethers are more useful than methyl ethers as they are usually solid at room temp. Reagents for the preparation of 2,4-dinitrophenyl ethers are chloro-2,4-dinitrobenzene[56] or fluoro-2,4-dinitrobenzene.[57–60] Characterization of phenols from microgram quantities by gas chromatography of their 2,4-dinitrophenyl ethers is possible.[61]

Different procedures for the preparation of these esters are described below, see recommended Methods *L*, *M*, *N* and *O*, p. 61. They may be characterized through paper, thin-layer, or gas chromatography. (See p. 108.)

### (*c*) *Trimethylsilyl Ethers*

Whereas methyl- and 2,4-dinitrophenyl ethers are used almost exclusively for the characterization of aromatically bound hydroxyl, trimethylsilyl ethers allow the characterization of aliphatic[62–65] as well as aromatic[66–68] bound hydroxyl. Trimethylsilyl ethers may be prepared[69] even from tertiary alcohols and highly hindered phenols. These ethers are recommended for the characterization of hydroxy compounds by gas chromatography, infrared spectroscopy, and mass spectroscopy.

The reagent is hexamethyldisilazane or a mixture of hexamethyldisilazane and chlorotrimethylsilane. The reaction is catalyzed by acids.

$$2ROH + (CH_3)_3Si{-}NH{-}Si(CH_3)_3 \rightarrow 2ROSi(CH_3)_3 + NH_3$$

$$ROH + (CH_3)_3SiCl \rightarrow ROSi(CH_3)_3 + HCl$$

Procedures for the preparation of these ethers are given below in recommended Methods *P*, *Q*, and *R*, pp. 61–62.

### (*d*) *Aryloxyacetic Acids*

Aryloxyacetic acids, prepared by the reaction of sodium phenate with sodium chloroacetate, are extremely convenient[70] both for the characterization of phenols through m.p. and for their alkalimetric estimation. The reaction is:

$$ArONa + ClCH_2COONa \rightarrow ArOCH_2COONa \xrightarrow{H^+} ArCH_2COOH$$

A procedure for the preparation of these derivatives is given in recommended Method *S*, p. 62.

The m.p.'s of many aryloxyacetic acids may be found in *Tables for the Identification of Organic Compounds*, The Chemical Rubber Company, Cleveland, Ohio.

(*e*) *Pseudo-saccharin Ethers*

Pseudo-saccharin chloride has been recommended by Meadoe and Reid[71] as a useful reagent for the identification of primary and secondary alcohols. More recently, Böhme and Opfer[72] have used this reagent for the identification of glycols and glycol derivatives. The reagent is prepared according to Jesurun[73] or to Böhme and Opfer,[72] who indicate two procedures for the preparation of pseudo-saccharin ethers.

(i) Dissolve in a flask (ground-glass), fitted with a reflux condenser, 0·3 g ethylene glycol and 0·8 g anhydrous pyridine in 10 ml pure chloroform. Add, dropwise, a solution of 2 g pseudo-saccharin chloride in 10 ml chloroform. When the vigorous reaction is over reflux the mixture until the evolution of hydrogen chloride has ceased. Allow to cool, filter off the precipitate by suction, wash several times with dilute hydrochloric acid, then with water, and finally recrystallize from glacial acetic acid. The reaction is:

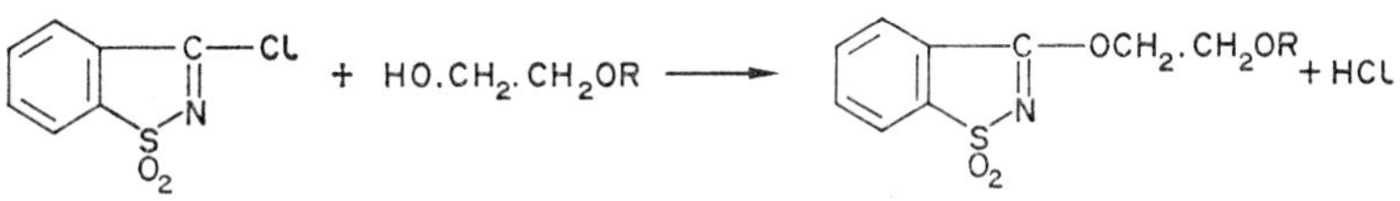


(ii) Mix in a flask (ground-glass), fitted with a reflux condenser, 0·76 g ethylene glycol monomethyl ether and 2 g pseudo-saccharin chloride. Fit the condenser with a calcium chloride tube and heat the flask at 120°C in an oil-bath for several hours until the evolution of hydrogen chloride has ceased. Allow to cool, dissolve the contents of the flask in chloroform, wash 2–3 times with water, then with saturated potassium hydrogen carbonate solution, and dry the chloroform solution over calcium chloride. When dry, evaporate the solvent and recrystallize the residue from ether-chloroform.

## 2. S-Alkylthiouronium Picrates

As mentioned above, esters and ethers of tertiary alcohols are difficult to prepare as the tertiary alcohols are usually dehydrated by the reagents used in esterification or etherification reactions. Schotte and Veibel[74] made use of the ready conversion of tertiary alcohols into tertiary alkyl chlorides or bromides (shaking of the alcohol with concentrated hydrochloric or hydrobromic acid at room temperature is sufficient) for the preparation of S-*tert.*-alkylthiouronium picrates which are readily crystallized. The alcohols may then be characterized through the m.p. of the picrates and their chromatographic behaviour. Furthermore, the equivalent weight of the alkylthiouronium picrate and consequently of the *tert.*-alcohol may be determined titanometrically or acidimetrically, see Part C, Chapter 10.8 (*b*), p. 91.

Jureček *et al.*[75] have shown that this method is useful for all alcohols which are readily converted into the corresponding chlorides, i.e. principally secondary and tertiary alcohols. They prefer, however, to isolate the S-alkylthiouronium-3,5-dinitrobenzoates instead of the picrates since the 3,5-dinitrobenzoates, in their opinion, permit a more accurate titanometric estimation of the equivalent weight than the picrates, which may easily be contaminated with picric acid. For the preparation see recommended Method *T*, p. 62.

### 3. N-Alkylbenzamides

Prajsnar,(76) making use of the ready conversion of tertiary and some secondary alcohols into alkenes, applied the Ritter reaction(77) for conversion of alkenes into N-alkylbenzamides by reaction with benzonitrile.

(*a*) *Tertiary Alcohols*

Add slowly 1 ml concentrated sulphuric acid and 1 ml benzonitrile to 1·5 ml of the alcohol. Shake occasionally, heat for 30 min at 60°C and then leave the mixture for a few h at room temp, cool by mixing with 15 g ice/water, stir, filter, wash with water, and recrystallize from 60% aq ethanol.

(*b*) *Secondary alcohols*

Add 2–3 ml concentrated sulphuric acid dropwise to 1·5 ml of the alcohol and 1 ml benzonitrile, shaking the flask during the addition. Keep the temperature below 60°C, then leave the mixture at room temp for a few h, pour into ice-water, stir, filter after 1–2 h and purify as above.

The reactions are:

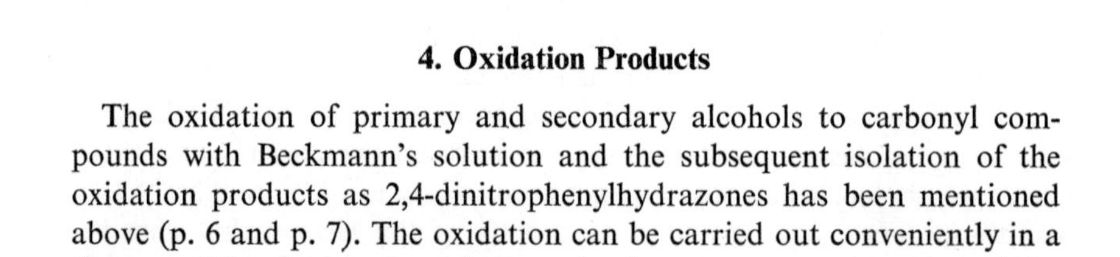

### 4. Oxidation Products

The oxidation of primary and secondary alcohols to carbonyl compounds with Beckmann's solution and the subsequent isolation of the oxidation products as 2,4-dinitrophenylhydrazones has been mentioned above (p. 6 and p. 7). The oxidation can be carried out conveniently in a Conway-dish, placing the substance in the outer compartment and the reagent in the inner one (Fig. 1).

Duke and Witman(78) recommend oxidizing the alcohols with potassium

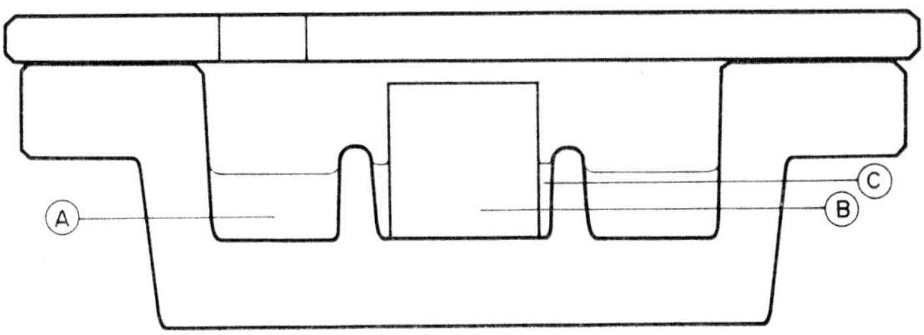


FIG. 1. Conway-dish. A. Sample + Beckmann-mixture; B. Glass plug used if only minute quantities of reagent are used; C. Reagent.

permanganate in 2 N sulphuric acid; any excess of potassium permanganate is destroyed with oxalic acid. The resulting carbonyl compounds are precipitated as 2,4-dinitrophenylhydrazones by adding 1% solution of 2,4-dinitrophenylhydrazine in 2 N perchloric acid.

Oxidation with xenon trioxide is mentioned in Part C, Chapter 10.5, p. 86 and Chapter 11.5 (*c*), p. 128.

## 5. Bromination Products

In aromatic compounds, hydrogen atoms in the position *ortho*- or *para*- to a hydroxyl group are readily substituted by bromine. The substitution is carried out by adding a potassium bromide–potassium bromate solution to an alkaline solution of the phenol, then acidifying the mixture. If the phenol is reasonably soluble in water, the bromine products may be obtained merely by adding bromine water to the aqueous solution or suspension of the phenol. In order to avoid the formation of derivatives of the "tribromophenol bromide" type (which may be recognized by their yellow colour and their ability to liberate iodine from an acidic solution of potassium iodide) not too great an excess of bromine should be added. The bromination products are usually insoluble in water and may be filtered off and recrystallized from aq ethanol.

Many heterocyclic hydroxy compounds behave analogously.

When *o*- and *p*-hydroxy-substituted benzoic acids are brominated the carboxyl group is easily replaced by bromine atoms. Tribromophenol is, therefore, formed by the bromination both of salicylic acid and of *p*-hydroxybenzoic acid.

CHAPTER 7

# CHROMATOGRAPHIC METHODS

## 1. Column Chromatography

This method is now seldom used, since the paper and thin-layer techniques are so much more rapid and convenient. For some coloured derivatives of hydroxy compounds, however, column chromatography has been applied to the separation of mixtures, e.g. of 4-dimethylamino-3,5-dinitrobenzoates (see p. 30). Jensen *et al.*[19] recommend alumina columns for the separation, and van Duin[20] used a nitromethane/silica gel column. Freytag and Baustian[21] found these columns unsatisfactory and recommended for partition chromatography a polyethylene column with heptane as stationary phase and dioxan/water (2/1) as the mobile phase. For adsorption chromatography an acid alumina column is sufficiently active, using heptane for elution.

Recently, Samuelson and Strömberg[79] described a method for the separation of polyols, carbonyl compounds, and sugars by partition chromatography on anion- and cation-exchange resins in ethanol–water mixtures. All polyols investigated were well separated on a cation-exchange resin in the lithium form, whereas certain mixtures with other solutes were better resolved on an anion-exchanger in its sulphate form.

With some complex mixtures overlapping occurred with both types of resin. Complete resolution of such mixtures can be achieved if fractions containing overlapping compounds are chromatographed on resin in another form.

## 2. Paper Chromatography

### (*a*) *Free Hydroxy Compounds*

Monohydric alcohols with less than about 10 carbon atoms are too volatile to be chromatographed in the usual way. According to Mukai *et al.*[80] unbranched aliphatic alcohols with 12–18 carbon atoms may be separated by reversed-phase paper chromatography, using paraffin or, better, kerosine (b.p. 185–215°C) as the stationary phase and aqueous acetic acid, preferably 80–85%, as the mobile phase.

Sherma and Thompson, Jr.[81] found it possible to separate normal alcohols from $C_7$ to $C_{12}$ by chromatography on Amberlite SA-2 ion exchange paper (containing about 45% by weight of Amberlite IR-120 cation-exchange resin in its sodium form). The chromatogram is developed, using the descending technique, with 8 M acetic acid or 75% methanol as mobile phase, the tank being equilibrated with the solvent for 1 h before starting the development. The spots are located by spraying with a vanadium oxinate solution (0·40 g of 8-hydroxyquinoline is dissolved in 25 ml each of xylene and glacial acetic acid and heated on a steam bath to 55°C; 0·20 g of ammonium metavanadate is then added with stirring and the resulting blue-black solution filtered through Whatman no. 42 filter after cooling. The reagent should be prepared fresh every 3 days.) Another spray is 0·50% solution of potassium permanganate in 0·1 M sulphuric acid. The dried chromatogram is drawn through this solution, and after the excess reagent has been allowed to drip off, the paper is steamed. Alcohol spots turn red, but impurities in the paper may give some colour.

Brownell[82] separated hydroxyl-containing substances by paper chromatography and located the spots by spraying a 6 × 22·5 in. sheet with 17 ml of 2% solution of 3,5-dinitrobenzoyl chloride in toluene and leaving the sprayed sheet overnight. Any hydroxyl groups present in the compound will have been esterified, whereas hydroxyl groups in the cellulose are not attacked. After steaming for 2 min the paper is drawn through 0·5% solution of sodium hydroxide (30 sec) and washed free of alkali with tap water. The wet sheet is then repeatedly drawn through a 0·002% aqueous solution of rhodamine and air-dried.

For polyhydric alcohols, different methods have been described for their paper chromatographic separation. Borecký and Gasparič[83] separate polyhydric alcohols by descending chromatography, using ethyl acetate saturated with water or ethyl acetate/ethanol/water (12/2/1) for developing the chromatogram, and spraying with a mixture of 9 ml of 5% silver nitrate and 1 ml ammonia to locate the spots. Finally the sheet is dried for 15–40 min at 100°C.

April[84] recommends ascending chromatography in a tank saturated with chloroform/ethanol vapour. The paper on which the solution and some test substances are applied is left in the tank overnight and then developed at 20°C ± 1°C with a mixture of isoamyl alcohol/ethanol/water (8/2/1). The developed chromatogram is dried overnight at room temp and the spots are then located by spraying with ammoniacal silver nitrate and drying at 100°C.

Other methods, mainly recommended for the separation of sugar alcohols, will not be mentioned here, as they will be discussed in the volume on carbohydrates.

The separation of phenols has been extensively studied. Macák and Kubeš[(85)] recommend paper impregnated with 20% ethanolic formamide or acetamide for the separation of the $C_7$–$C_{14}$ alkylphenols. After drying the paper, benzene solutions of the phenols are applied and the chromatogram is developed at 15–17°C with cyclohexane, either by the ascending or, better, by the descending technique. The spots are located by spraying with Pauly's reagent. Formamide-impregnated paper can be used for all the phenols studied, and acetamide-impregnated paper for phenol and the $C_7$–$C_9$ alkylphenols only.

Gasparič *et al.*[(86)] studied the separation of alkylphenols on paper impregnated with a 25–50% benzene solution of dimethylformamide or a 10% solution of paraffin in hexane. The impregnated paper is suspended for 5–15 min with the starting line at the bottom and the sample is then applied as a solution (0·5–2%) in benzene or ethanol. As the mobile phase (descending technique) hexane is used for dimethylformamide-impregnated paper, and methanol/water for paraffin-impregnated paper. The spots are located by spraying with iron(III) chloride (blue spots on light yellow background) or other sprays commonly used for locating phenols.

For the lower alkylphenols dimethylformamide-impregnated paper is best, but for the higher alkylphenols paraffin-impregnated paper is preferable.

Locke and Sherma[(87)] used Amberlite SA-2 ion exchange paper in the same way as Sherma and Thompson, Jr.[(81)] did for alcohols (see p. 44). The chromatogram is developed by the descending technique with acetic acid/water (1–8 M solution). After drying, the spots are located by spraying with a mixture of equal volumes of 5% iron(III) chloride and 1% potassium hexacyanoferrate(III). The paper is steamed for 5 min, after which blue spots develop.

Clark[(88)] used Reeve–Angel Grade SB-2 Amberlite IR A-400 ion exchange loaded paper and cyclohexanol/ethyl acetate/acetic acid (5/1/1) for developing the chromatogram (descending technique). After drying the paper in a circulation air-oven at 50°C for 5 min the spots were located by spraying with diazotized sulphanilic acid followed by 20% sodium carbonate solution.

Fujiwara and Sugimoto[89] obtained good separation of alkyl substituted phenols by using paper impregnated with dibutyl adipate and developing the chromatogram with aqueous ammonia. In general, an increase in the number of carbon atoms in the side-chain decreases the $R_f$ value; *m*-monoalkylphenols have the highest $R_f$ values; the *o*-isomers have the lowest. For dialkylphenols the order is found to be 3,5- > 3,4- > 2,5- > 2,3- > 2,4- > 2,6-substitution.

Aarna and Karjama[90] recommended a circular paper chromatographic method for the separation of 27 different monohydric and dihydric phenols. A sheet of paper pretreated with ethanol and then with a 20% ethanolic solution of formamide is placed between two glass plates, the upper one with a hole at the centre. The mixture of phenols is applied to the paper through the hole, followed by a mixture of benzene/cyclohexane (1/10). Only monohydric phenols move with this solvent, whereas dihydric phenols remain at the point of application. The central circle is then cut out, and a hole is made in another sheet of paper so as to fit exactly the disc cut out. This new sheet of paper is then placed between the glass plates and developed with 5% acetic acid. The phenols are located by spraying with a solution of 0·1 g diazotized sulphanilic acid in 10 ml sodium carbonate solution.

Micheel and Leifels[91] treated cellulose paper with ethyleneimine, causing an inclusion of polyethyleneimine. The procedure is rather complicated and should be studied *loc. cit.* and in a preceding paper by Micheel and Schminke.[92] Chromatography on paper impregnated in this way permits the separation of nitrophenols [coloured spots when developed with butanol/glacial acetic acid/water (4/1/5)], nitroaminophenols [coloured spots when developed with butanol/pyridine/water (3/1/1)], and dihydroxybenzoic acids [colourless spots when developed with butanol/acetic acid/water (4/1/5) or butan-2-ol/formic acid/water (75/15/10)], the spots becoming visible by spraying with *p*-nitrobenzenediazonium borofluorate.

Gumprecht[93] has made a thorough study of developers for the separation of 59 different phenols, recommending 3 polar developers [water/dioxan/glacial acetic acid (75/24/1), water/methanol/glacial acetic acid (75/24/1) and water] and three non-polar developers (benzene saturated with water, toluene saturated with water, and mineral spirit saturated with water). Spots are located by spraying with cerium(III) ammonium nitrate.

*(b) Derivatives of Hydroxy Compounds*

(i) *Potassium alkyl xanthates.* Several authors [94–97] have made use of the facile formation of potassium alkyl xanthates for the chromatographic separation of primary and secondary alcohols. Convenient and nearly identical methods were recommended by Gasparič and Borecký[96] and by Pohloudek-Fabini and Beyrich.[95] Some drops of the sample are mixed with 0·1 g of pulverized potassium hydroxide, the mixture warmed to dissolve the potassium hydroxide and then 0·5–1 ml of carbon disulphide is added. Stir the mixture with a glass rod for some min, then evaporate the excess carbon disulphide by heating the test-tube in hot water. Dissolve the residue in a few drops of water (or, for higher alcohols, acetone) and apply the solution to paper impregnated with 10% potassium carbonate solution or with phosphate buffer, pH 12. Develop the chromatogram with butanol saturated with 30% potassium carbonate solution or with propanol/water mixture (97/3) or (95/5), using the ascending or descending technique. Localize the spots by illumination in ultraviolet light, or by spraying with 10% nickel sulphate solution (yellow to orange spots on white background) or with 1% ammonium molybdate containing 2% of sulphuric acid (blue spots).

Pozetti[97] recommended this reaction for a spot test for primary, secondary, and tertiary alcohols. Add to each spot on a chromatogram of a mixture expected to contain alcohols one drop of carbon disulphide and a few centigrams of pulverized potassium hydroxide. Spray after 2 min with 2% ammonium molybdate acidified with sulphuric acid and then with chloroform. After drying for 5 min at 100–110°C all alcohols tested gave positive results with colours varying from blue to violet.

Berbalk[98] and later Frahn[99] made use of the xanthates for paper-electrophoretic identification of alcohols. Prepare the xanthates as above, dissolve in water, and apply the aqueous solution to Schleicher and Schüll 2043 b chromatography paper saturated with veronal buffer (pH 9·8)[98] or sodium hydrogen carbonate solution[99] as electrolyte and carry through the electrophoresis for 30–60 min at 4°C with a voltage of 20–25 V/cm. Dry the paper at 100°C and locate the spots by illuminating with u.v. light. Twenty-six alcohols including four glycols, benzyl alcohol, $\alpha$- and $\beta$-phenylethyl alcohol, and furfuryl alcohol have been tested. The displacement for methylxanthate is 6–7 cm, and the displacements for other xanthates were 49–87% of this value.

(ii) *(Substituted) phenylurethanes.* Attaway[100] studied different substituted phenylurethanes and found *m*-nitrophenylurethanes best for the chromatographic identification of $C_1$–$C_8$ saturated alcohols, and *o*-nitrophenylurethanes best for $C_6$–$C_{12}$ alcohols, using the descending technique. For $C_1$–$C_8$ alcohols the sheet of paper was soaked with a methanol/propylene glycol mixture and dried for 1 h. The chromatogram was developed with Skellysolve C saturated with methanol, or with a methanol/propylene glycol mixture (80/20); for $C_6$–$C_{12}$ alcohols the paper is soaked with 7% or 10% solution of vaseline in Skellysolve C and air-dried for 15 min. The chromatogram was developed with methanol/water (95/5). The spots were located by illumination in u.v. light (360 nm). For terpene alcohols, Attaway *et al.*[(101)] recommended *o*-nitrophenylurethanes or *p*-nitrophenylazophenylurethanes, chromatographed on paper impregnated with vaseline and developed with methanol/water (90/10).

(iii) *p-Nitrobenzoates.* These esters, prepared as indicated (p. 29), can be separated according to Komárek *et al.*[(13)] by descending paper chromatography. For esters of $C_1$–$C_7$ alcohols, Whatman no. 2 paper impregnated with 70% formamide in ethanol and developed with cyclohexanol was used. For $C_7$–$C_{12}$ alcohols, Whatman no. 2 paper impregnated with 3% paraffin oil in cyclohexane and developed with formamide/methanol/water (3/2/1) was used, and for $C_{10}$–$C_{16}$ alcohols Whatman no. 2 paper impregnated with 5% paraffin oil in cyclohexane and developed with formamide/methanol/water (9/5/1) was used. The spots were located by spraying with tin(II) chloride and then with Ehrlich's reagent.

(iv) *3,5-Dinitrobenzoates.* Micheel and Schminke[(102)] separated the 3,5-dinitrobenzoates of $C_1$–$C_{12}$ alcohols by chromatography on acetyl-cellulose paper, developing the chromatogram with ethyl acetate/dioxan/water (20/45/46) and locating the spots by illumination in u.v. light. Gasparič and Borecký[(103)] have studied the chromatographic separation of 150 3,5-dinitrobenzoates of alcohols, glycols, polyethylene glycols, and phenols. They located the spots by first spraying with a solution of tin(II) chloride, thus reducing the nitro groups to amino groups, and then with a solution of *p*-dimethylaminobenzaldehyde with which the amino groups form coloured Schiff bases. They found it possible to detect amounts down to 1 μg of 3,5-dinitrobenzoate.

Gauthier and Mangeney[(104)] have also shown that polyethylene glycols may be characterized through paper chromatography of their 3,5-dinitrobenzoates.

(v) (*Substituted*) *phenylazobenzoates.* It was indicated previously (p. 30) that Pohloudek-Fabini and Münchow[24] recommended phenylazo-4-benzoates for the characterization of hydroxy compounds. The preparation and paper chromatographic separation is described below as recommended Method *E* (p. 58).

Churáček[105] has shown that 4′-dimethylaminoazo-4-benzoates are extremely well suited for paper chromatographic separation of hydroxy compounds. The reagent, 4′-dimethylaminoazo-4-benzoyl chloride, is not hydrolyzed by the humidity of the air; the esters are easily prepared by conventional methods and easily separated by chromatography on paper impregnated with paraffin oil, using dimethylformamide/methanol/water (8/2/1) as the mobile phase. Spraying is unnecessary as the spots are directly visible.

(vi) *Other esters.* Jackson and Hayward[106] described paper chromatography of nitrates and *p*-toluenesulphonates of polyhydroxy compounds, and Tulus and Güran[107] that of 2,4-dinitrophenylsulphenates, prepared as indicated (p. 33). They recommended paper impregnated with a mineral oil [fraction b.p. 176–210°C (20 % solution in acetone for $C_1$–$C_5$ alcohols, 10 % solution in acetone for $C_5$–$C_8$ alcohols)], and developed the chromatograms (descending technique) with dioxan/water/acetone (60/45/75) for $C_1$–$C_5$ alcohols, and (60/45/60) for $C_5$–$C_8$ alcohols.

Maruta *et al.*[108] prepared esters of $C_2$–$C_{18}$ alcohols with crotonic acid and added on mercury(II) acetate to these unsaturated esters, separating the addition compounds by reverse-phase chromatography, and locating the spots by spraying with a solution of diphenylcarbazone (purple spots). An estimation of the amount of alcohol can be made by measuring the areas of the spots.

(vii) *Azo compounds.* Crump[109] and Makes[110] recommended coupling with diazotized aromatic amines, preferably *p*-nitroaniline, for the identification of aromatic hydroxy compounds. The azo dyes are extracted from acid solution with ether, the ethereal layer dried over magnesium sulphate, the solvent removed by evaporation, and the residue dissolved in chloroform. This solution is applied to paper impregnated with formamide (20 % solution in acetone) or dimethylformamide (30 % solution in benzene) and the chromatogram developed with benzene/cyclohexane/dipropylene glycol (30/70/3) or cyclohexane/pyridine (10/1) by the ascending technique. The spots are directly visible, but spraying with 5 % sodium carbonate solution increases the intensity of the colours.

(viii) *Ethers.* Churáček *et al.*[111] indicated a method for the microscale preparation of alkyl 2,4-dinitrobenzyl ethers in sealed vials. Introduce into a micro test-tube 0·1 g of 2,4-dinitrobenzyl bromide and 0·1 ml of an alcohol (for alcohols with more than 5 carbon atoms a few drops of benzene are added). Seal the tube, and heat it on a water bath for 2 h ($C_1$–$C_4$ alcohols) or 4 h ($C_4$–$C_8$ alcohols). For alcohols above $C_8$ heat for 4 h in an oil-bath at 160°C. Cool, open the vial, dilute the reaction mixture with benzene. Use this solution without purification for paper chromatography; for $C_1$–$C_7$ alcohols use Whatman no. 2 paper and develop with dimethylformamide/cyclohexane (1/1) or dimethylformamide/(cyclohexane/benzene 25/1) (1/1) mixture; for $C_8$–$C_{12}$ alcohols, use Whatman no. 2 paper impregnated with 3% paraffin oil and develop with dimethylformamide/methanol/water (3/2/1). Locate the spots by spraying with a freshly prepared mixture of equal volumes of sodium ethoxide (40 g sodium in 500 ml ethanol) and acetone: green spots that turn brown after a time are given. Any excess reagent also gives a green spot, changing quickly to a violet hue, and eventually becoming brown. Down to 0·8–1·2 μg can be detected by spraying with Ehrlich's reagent.

The $R_M$-values of alcohols vary linearly with the number of carbon atoms, isomeric alcohols differing only negligibly in their $R_f$-values.

(ix) *S-Alkylthiouronium iodides.* Sucker[(112)] recommended conversion of alcohols to alkyl iodides and, by reacting the alkyl iodide with thiourea, to S-alkylthiouronium iodides. For the conversion of an alcohol to the corresponding iodide, Stone and Shechter[(113)] refluxed a mixture of 1 mole of the alcohol, 2 moles potassium iodide and 3 moles 95% phosphoric acid with stirring for 5 h (for diols the molar ratio is 1/4/6 and the mixture is heated for 5 h at 100–120°C). After cooling, 150 ml water and 250 ml diethyl ether are added, then the ethereal layer is separated, the iodine liberated is removed by washing with a solution of sodium thiosulphate, then with a saturated solution of sodium chloride, and dried over sodium sulphate. After removal of the ether by evaporation, the residue is fractionated on an effective column.

A procedure for the conversion of the alkyl iodide to S-alkylthiouronium iodide is described below (recommended Method *T*, p. 62).

## 3. Thin-layer Chromatography

### (*a*) *Free Hydroxy Compounds*

Alcohols may be separated by chromatography on Silica gel G-plates

(Subbarao *et al.*[114]) or on Kieselguhr G-plates (Purdy and Truter,[115] Nobuhara[116]), using ether/petroleum ether,[114] cyclohexane,[115] or acetone/water (60/40) or dioxan/water (60/40)[116] for developing the chromatogram. The spots were located by spraying with 50% sulphuric acid[114] or concentrated sulphuric acid,[115] followed by heating at 100–160°C for 10–15 min (brown spots on white background) or spraying with Rhodamin 6 G and illuminating in u.v. light (saturated and mono-enolic alcohols gave yellowish-orange spots on a yellowish background, dienolic alcohols gave violet-blue spots).[116]

Bandyopadhyay and Chakrabarty[117] found that Kieselguhr G-plates impregnated with a 10% solution of liquid paraffin in petroleum ether and then air-dried for 15 min allowed a better separation than unimpregnated plates. The chromatogram was developed with acetone/water (75/25) after saturation of 80% of the developer with liquid paraffin. The spots are located by spraying with 1% ethanolic molybdophosphoric acid. After heating for 10–15 min at 110°C saturated fatty alcohols gave white spots, and sperm-oil alcohols gave bluish-violet spots against a greenish background.

McSweeney[118] used a similar technique to separate terpene alcohols, developing the chromatogram with acetone/water/liquid paraffin (65/35/0·5) and locating the spots in iodine vapour. Also good separation on silica gel plates was obtained with benzene/ethyl acetate (95/5) for developing the chromatogram.

Stahl and Vollmann[119] recommended for the separation of terpene- and sesquiterpene alcohols chromatography on plates coated with kieselguhr containing 3% silver nitrate. The plates were developed in tanks saturated with the vapour from methylene chloride/chloroform/ethyl acetate/propanol (50/50/5/5) or methylene chloride/chloroform (40/60). After spraying with a 20% solution of molybdophosphoric acid in methanol and heating for 5–10 min at 110°C the alcohols were located as blue spots on a yellow background. Terpenes, sesquiterpenes, and diterpenes formed 3 distinct groups, whereas no separation was obtained on silica gel plates.

For the separation of polyols, Knappe *et al.*[120] recommended plates coated with alumina (chloroform/toluene/formic acid (80/17/3) for developing), kieselguhr G (butanol saturated with 1·5 N ammonium hydroxide for developing) or Kieselguhr G with Ultramid 1 C (6/1) (chloroform for developing). They have studied 17 diols and triols.

Fischer and Koch[121] indicated that *erythro*- and *threo*-forms can be separated on kieselguhr-coated plates with methanol/acetone (1/1), water saturated with chloroform, water saturated with benzene, or water saturated with tetrachloromethane used for developing the chromatogram.

de Simone and Vicedomini(122) have observed that, in alkaline medium, complex formation takes place between lead(II) ions and polyols containing at least two adjacent hydroxyl groups.

They coated plates with a slurry of 30 g Silica gel H (Merck) and 70 ml 0·1 M lead nitrate, using plates without lead nitrate for comparison. All plates were activated at 110°C for 30 min. The chromatograms were developed with ethanol saturated with gaseous ammonia or ethanol/32% ammonia/water (23·4/2·2/1) and the spots were located by spraying with a solution of potassium permanganate containing sodium carbonate.

Akhrem *et al.*(123) separated acetylenic alcohols and glycols on alumina-coated plates, developing the chromatograms with diethyl ether/benzene and locating the spots in iodine vapour.

For thin-layer chromatography of phenols different coatings have been suggested. Lipina(124) and Vazquez and Montes(125) used silica gel mixed with 5–10% of starch and activated the plates by heating for 1–2 h at 100–110°C. Chromatograms were developed with a mixture (7/3) of a petroleum fraction (b.p. 61–69°C) and ethyl acetate(125) or with benzene/acetic acid/water (2/2/1)(124) and the spots were located by spraying with diazotized *p*-nitroaniline. Tadema and Batelaan(126) found it possible to detect sub-micro quantities of halogeno- and nitrophenols by exposing the developed chromatogram to bromine vapours, then spraying with aqueous potassium iodide.

Dittmann(127) and Chafetz *et al.*(128) preferred plates coated with cellulose powder activated for 4 h at 105°C and they developed the chromatogram with amyl alcohol/formic acid/water (40/40/2) or propanol/water (63/35). Spots were located by spraying with diazotized sulphanilic acid.

Van Sumere *et al.*(129) used a mixture of equal amounts of silica gel and cellulose powder for coating the plates, thus combining the advantages of thin-layer and paper chromatography. The plates were steamed just before use. The chromatograms were developed with toluene/ethyl formate/formic acid (5/4/1) or chloroform/acetic acid/water (4/1/1). The spots were located by illumination in u.v. light or by spraying with diazotized *p*-nitroaniline followed by 5% sodium carbonate.

For separating *m*-dihydric phenols from *o*- and *p*-dihydric phenols Aly(130) used plates coated with silica gel containing 4–5% of silver nitrate.

The *o*- and *p*-dihydric phenols were oxidized to the corresponding quinones, whereas the *m*-compounds remained unoxidized. The chromatogram was developed, e.g. with benzene/methanol/acetic acid (8/1/1) or benzene/acetic acid (8/2). To prevent discoloration of the plates before use, they were stored in a desiccator in darkness.

Bark and Graham[(131)] and Graham[(132)] found plates coated with cellulose impregnated with polyamide (Nylon 66) convenient for the separation of substituted phenols. Prepare a slurry of 11 ml of a solution of polyamide in formic acid, 13·5 g cellulose powder, and 74 ml formic acid, coat the plates to a thickness of 0·3 mm, air-dry for 24 h at 25°C, activate for 15 min at 80°C and cool the activated plates in an evacuated desiccator.

The chromatograms were developed with a mixture of cyclohexane/acetic acid (93/7) or with 10% aqueous acetic acid, the spots were located by spraying with alkaline potassium permanganate (yellow spots on purple background). Altogether 178 phenols were studied.

$C_1$–$C_{12}$ alkylphenols were separated on polyamide-coated plates (Diamond[(133)]), using water/dimethylformamide/formamide (60/10/10) or N sodium hydroxide/methanol (70/30) for developing the chromatogram and 0·5% aqueous Fast Blue Salt B solution for locating the spots (brown spots on white background).

Reviews of materials for coating the plates and mixtures for developing the chromatograms were given by Crabtree and McGill[(134)] and by Parihar *et al.*[(135)] Van Dijk and Mijs[(136)] pointed out that methanol-containing development systems may cause irreproducible results as even a minute amount of methanol vapours in the tank will be strongly adsorbed to the dry parts of the coating, thus modifying in an uncontrolled way the adsorption of the phenols.

(*b*) *Derivatives of Hydroxy Compounds*

(i) *Potassium alkyl xanthates* of lower alcohols were recommended by Röhr and Chiari[(137)] for thin-layer chromatography on cellulose-coated plates; propanol/ligroin (b.p. 100–150°C)/conc. ammonia (60/30/20 or 60/20/20) were used as eluents. The spots were located by spraying with 10% nickel sulphate (yellow or orange spots) or a mixture of equal volumes of 2 N sulphuric acid and 1% ammonium molybdate solution (violet spots, fading after some hours).

(ii) (*Substituted*) *phenylurethanes* were used by Derbesy and Naudet[(138)] for the separation of diols by chromatography on silica gel coated plates, washed with ether and reactivated, developing the chromatogram with ether/hexane/glacial acetic acid (84/15/1) and locating the spots with a spray of chromic acid in sulphuric acid, followed by heating to 180°C for 30 min.

(iii) *Acetates.* Dallas[139] recommended as a method for identification of polyglycols an acetylation followed by thin-layer chromatography on plates coated with a 1/1 mixture of Silica gel G and Kieselguhr G dried at 110°C. Dissolve 1–10 mg sample in 100 ml pyridine. Apply 1 μl to the plate. Mix 20 ml acetic anhydride into the pyridine solution and again apply 1 μl to the plate. Heat the mixture just to the boiling point, cool, and apply a third 1 μl sample to the plate. Keep the solution close to the boiling point for several min, cool, and apply a final 1 μl sample. Develop the chromatogram in a paper-lined tank with mixtures of cyclohexane/ether/ethanol/formic acid (30/82/6/2) for glycerol and 1,1′-diglycerol, (23/67/15/15) for linear triglycerol, and (10/102/6/2) for cyclic triglycerol. Spray with 0·5 g thymol dissolved in a mixture of 5 ml concentrated sulphuric acid and 95 ml ethanol, heat to 120°C for 20 min to give permanent grey-purple spots on a pale mauve background.

(iv) *3,5-Dinitrobenzoates.* Dhont and de Rooy[140] used Silica gel G, Dhont and Dijkman[141] used kieselguhr impregnated with ethyleneglycol monophenyl ether for thin-layer chromatography of alkyl 3,5-dinitrobenzoates. The silica gel plates were developed with benzene/petroleum ether b.p. 60–80°C (1/1) or with hexane containing 15 or 25% of ethyl acetate; the kieselguhr plates were given three successive developments with petroleum ether. The spots (yellow to orange) were located by spraying with a solution of 1-naphthylamine.

Mehlitz *et al.* [142] differentiated between 3,5-dinitrobenzoates of saturated and of unsaturated alcohols by developing the chromatogram with isopropyl ether/petroleum ether b.p. 40–60°C (12/88 for $C_1$–$C_8$ alcohols, 6/94 for higher alcohols). The spray was 0·05% fluorescein in 0·1 N sodium hydroxide. On exposing the sprayed plate for a short time (less than 1 sec) to a dilute bromine atmosphere, the esters of unsaturated alcohols gave yellowish spots while those of saturated alcohols gave a reddish colour.

Severin[143] recommended two-dimensional chromatography of the 3,5-dinitrobenzoates of $C_1$–$C_{14}$ alcohols, first on kieselguhr-coated plates with cyclohexane/ethyl acetate (90/10); then, after having dried the plates, impregnated them with a 10% solution of paraffin oil (density 0·865–0·890) in petroleum ether b.p. 30–50°C and removed the solvent by evaporation, the plates were turned 90° and developed with glacial acetic acid/water. Spots were located by spraying with 0·1% Rhodamin B in ethanol, giving violet spots on a pink background.

(v) *Nitrates* were recommended by Malins *et al.*[144] for the identification of hydroxy compounds as they are easily prepared in test-tubes by reaction with acetyl nitrate. The crude reaction mixture is then chromatographed on Silica gel G coated plates and developed with hexane in a tank lined with filter paper. Spots are located, after spraying with a solution of 2′,7′-dichlorofluorescein, by illuminating the plate in u.v. light or by exposing it to iodine vapour. The unique chromatographic and spectral properties of nitrates make them more easily separated and analysed than the corresponding alcohols.

(vi) *β-Alkoxypropionitriles.* Buzlanova *et al.*[145] converted alcohols to β-alkoxypropionitriles by cyanoethylation, reacting them with acrylonitrile:

$$ROH + CH_2{=}CHCN \rightarrow ROCH_2CH_2CN$$

The nitriles were then separated by thin-layer chromatography on silica gel coated plates. The chromatograms were developed with cyclohexane/ethanol/heptane (10/7/3) for $C_1$–$C_6$ alcohols or tetrachloromethane/ethyl acetate (20/5) for $C_7$–$C_{12}$ alcohols and the spots were located by spraying with a mixture of equal volumes of 0·2% ethanolic methyl red and 0·1% ethanolic methyl blue.

There is a linear dependence between the number of carbon atoms of an alcohol and the $R_f$-value of its β-alkoxypropionitrile.

(vii) *2,4-Dinitrophenyl ethers* are well suited, according to Wildenhain and Henseke[146] and to Parihar *et al.*,[147] for the thin-layer chromatographic separation and characterization of phenols on silica gel or magnesium silicate coated plates[146] or on plates coated with alumina containing 20% of calcium sulphate,[147] developing the chromatograms with, e.g., benzene/chloroform (1/1) or cyclohexane/cyclohexanol/glacial acetic acid (100/10/4), the tank being saturated with vapour of the developer system. The spots were located by spraying with propylamine (lemon-yellow spots). Heating at 130°C for 15 min makes the spots more visible. $R_f$-values for ethers of monohydric phenols were 0·60–0·95, for dihydric phenols $R_f$ = 0·28–0·73, and for trihydric phenols $R_f$ = 0·05–0·64.

(viii) *Trimethylsilyl ethers.* Leibman and Ortiz[148] converted glycols into trimethylsilyl ethers by treating them with 60 μl hexamethyldisilazane/

chlorotrimethylsilane (2/1) and 0·14 ml pyridine, chromatographing the mixture on plates coated with silica gel containing a fluorescing indicator, and developing the chromatogram with heptane. The spots were located by illuminating the plate with u.v. light of 254 nm.

### 4. Gas Chromatography

Gas chromatography has become one of the most powerful methods for the characterization and estimation of hydroxy compounds. As most of the methods recommended have been used for the estimation of hydroxy compounds, they will be discussed in Part C.

CHAPTER 8

# RECOMMENDED METHODS FOR THE PREPARATION OF DERIVATIVES OF HYDROXY COMPOUNDS

## 1. 3,5-Dinitrobenzoates

### *Method A* (*Saunders et al.*[(15)])

Mix 0·5 g 3,5-dinitrobenzoic acid and 1 g phosphorus pentachloride in a test-tube and warm until liquid. A spontaneous reaction takes place which is maintained by gentle heating for 5 min, the reaction being vigorous during this period. Pour the liquid reaction mixture onto a watch glass and transfer it after crystallization to a porous plate and remove the phosphoryl chloride by rubbing.

Add 0·5 g of the crude 3,5-dinitrobenzoyl chloride to 1 ml of the alcohol and reflux the mixture for 5 min. Add 10 ml water and cool the solution in an ice-bath, initiating the crystallization by scratching. Isolate the ester by filtration, wash with 10 ml 2% sodium carbonate and recrystallize from acetone or (aqueous) ethanol. Ethanol may, however, effect transesterification of the product.

### *Method B* (*Dunbar and Ferrin*[(16)])

Treat 1 ml of the alcohol (or 1 g of a solid alcohol dissolved in ether) with metallic sodium until no further reaction can be observed. If spontaneous reaction does not occur, heat moderately to start and maintain the reaction. Then add 1–2 drops of the alcohol to remove any excess of sodium, followed by 0·5 g 3,5-dinitrobenzoyl chloride, moderating, if necessary, the reaction by cooling in order to avoid charring. Add a little more 3,5-dinitrobenzoyl chloride if the reaction mixture remains purple. The esterification is finished when only a slight colour is left. Shake vigorously with 10 ml water, decant the water and purify the residue as above.

### *Method C* (*Huth*[(17)])

Mix 10 ml of 10% aqueous solution of the hydroxy compound with 10 ml acetone and add sufficient potassium carbonate to produce only one visible layer at the end of 15 min (up to 30 g potassium carbonate may be necessary). Decant the layer, add to it 0·5 g 3,5-dinitrobenzoyl chloride and reflux for 10 min. Distil most of the acetone off, and recrystallize the 3,5-dinitrobenzoate as above.

Even tertiary alcohols may be esterified in this way when pyridine instead of acetone is used as solvent.

## 2. 4-Dimethylamino-3,5-dinitrobenzoates

### *Method D* (*van Duin*[20])

Prepare 4-dimethylamino-3,5-dinitrobenzoyl chloride from 4-chlorobenzoic acid through 4-chloro-3,5-dinitrobenzoic acid and 4-dimethylamino-3,5-dinitrobenzoic acid, according to Jensen *et al.*[19]

Mix a 1–5% solution of the alcohol in anhydrous benzene with the equivalent amount of a 1% solution of the acid chloride in anhydrous benzene. Add a slight excess of pyridine, and leave the mixture to stand for 24 h at room temp or for 1 h at 80°C (reflux). The ester separates out on cooling, in some cases after concentration of the solution by evaporating some of the solvent.

## 3. Esters of Azobenzene-4-carboxylic Acid

### *Method E* (*Pohloudek-Fabini et al.*[22–24])

Mix in a test-tube 15–20 mg of the sample, 70 mg azobenzene-4-carbonyl chloride, 0·5 ml benzene and 0·05 ml pyridine. Reflux for 4 h, cool, mix with 2 drops of saturated sodium carbonate solution, dilute to 10 ml with pyridine, and chromatograph a 0·01 ml aliquot on paraffin-impregnated Schleicher and Schüll no. 2230 paper using the ascending technique; the mobile phase is dimethylformamide/methanol/water (4/1/0·65) or methanol/water/heptane (100/72/25 wt/wt/vol).

To impregnate the paper soak it with a 5% solution of liquid paraffin in toluene and remove the toluene by leaving the soaked sheet suspended for several h.

$C_1$–$C_{12}$ alcohols can be separated by this procedure. For estimation cut out the spots and elute with heptane. Measure the absorbance at 322 nm, this absorption being due to the azo-group and thus independent of the alkyl moiety. The maximum error is $\pm 10\%$ for a single alcohol, $\pm 15\%$ for multicomponent systems.

## 4. Esters of 4′-Nitroazobenzene-4-carboxylic Acid

### *Method F* (*Bosvik et al.*[26])

Prepare esters of 4′-nitroazobenzene-4-carboxylic acid by reaction of the hydroxy compound with the corresponding acid chloride (which is prepared according to Hecker).[25] Chromatograph on Schleicher and Schüll no. 2045 bM paper impregnated with dimethylformamide (by immersing it for 20 min in a 30% solution in acetone and afterwards drying it by pressing it between filter paper), using the ascending technique with 2,2,4-trimethylpentane as mobile phase, the tank being saturated with vapours of both phases. Cut out the spots and elute with acetone. For i.r. spectrometry make potassium bromide discs from 1 mg sample and 300 mg potassium bromide; for X-ray diffraction patterns mix 0·1–0·2 mg with silica (for calibration) and mount the mixture on the cellulose tape. Spectra are taken with an X-ray diffraction camera, using Cu $K\alpha$-radiation.

### *Method G* (*Amin and Samad*[27])

*Monohydric phenols.* Dissolve 0·5 m-equiv in 2 ml dry pyridine and add 0·725 m-mole (0·21 g) of 4′-nitroazobenzene-4-carbonyl chloride dissolved in dry benzene. Stopper the flask, and heat it for 2 h on the steam-bath. Dilute with 5 ml dry pyridine and 10 ml of 5% solution of pyridine in benzene. Filter through a column of 10 g of activated alumina wetted with the 5% pyridine/benzene solution. Elute the ester from the column with 30 ml of this solution. Remove the solvents, and recrystallize the ester from ethanol/benzene (1/1).

*Polyhydric phenols.* Dissolve 0·5 m-mole of the sample in 50 ml of freshly distilled ether, add 0·725 m-mole reagent and 1·5 ml pyridine. Keep the flask at room temp for 24 h, then add more reagent (0·725 m-mole for dihydric, 1·45 m-mole for trihydric phenols) and leave the flask for another 24 h. Add 2 ml pyridine and reflux for 6 h. Remove the solvents by distillation and dry the residue thoroughly. Extract it then with 250 ml benzene, wash the extract with 20% sulphuric acid, filter, wash with water, dry over sodium sulphate and filter through a column of alumina. Elute the main lower red band with benzene and the upper band with acetone. Recrystallize from acetone/benzene (1/1). The lower band contains diesters of dihydric phenols and triesters of trihydric phenols, the upper band monoesters of dihydric and mono- and diesters of trihydric phenols.

For separation, prepare a column as follows: A 4/1 (vol/vol) mixture of nitromethane and nonane separates in two layers. Half-fill a tube, 20 cm × 2 cm, with the nitromethane layer. Make a sludge of 20 ml of the nitromethane layer with 5 g of kieselguhr impregnated with dimethyloxosilane and 5 g of the nonane layer and add it to the column. After draining off the column add 2–3 mg of the esters dissolved in the nitromethane layer and develop the chromatogram with the nitromethane layer under reduced pressure. Elute the bands with acetone.

## 5. Acid Esters of Phthalic or Tetrachlorophthalic Acid

### *Method H* (*Fessler and Shriner*[(30)])

Prepare a solution of ethylmagnesium bromide from 0·75 ml ethyl bromide in 25 ml anhydrous ether and 0·24 g magnesium. Add 0·01 mole of the alcohol dissolved in 10 ml anhydrous ether. When the evolution of ethane stops add anhydrous dioxan followed by 0·01 mole of phthalic or tetrachlorophthalic anhydride and leave the flask overnight in the refrigerator. Then treat the reaction mixture with 50 g ice and add 12·5 ml 2 N hydrochloric acid. Remove unreacted acid anhydride or acid by filtration, separate the ethereal layer and extract the aqueous layer with 15 ml ether. Wash the combined ethereal layers twice with 10 ml and 5 ml respectively of ice-cold 5% sodium hydroxide solution. Cool the combined alkaline extracts in an ice-bath and acidify with 6 N hydrochloric acid. Isolate the ester by filtration and purify it by recrystallization from petroleum ether or ligroin.

## 6. Esters of 3-Cyano-3,3-diphenylpropionic Acid

### *Method I* (*Salmon-Legagneur and Neveu*[(39)])

Mix 0·2–0·5 g of the alcohol with the calculated amount of 3-cyano-3,3-diphenylpropionyl chloride (prepared according to Salmon-Legagneur[(40)]); heat the mixture for 10–15 min on a water-bath at 60–80°C (if less than 0·2 g of the alcohol is available, add 1–2 drops dioxan). Cool and add some ml of dilute sodium carbonate and leave the mixture to stand for several h at room temp to remove excess acid or acid chloride. Filter, wash the crystals with water, and recrystallize from ethanol or aqueous (60%) ethanol.

Tertiary alcohols are converted into $R_3COMgBr$ as in Method *H*, and the alcoholate is then refluxed for 1 h with a dioxan solution of the reagent. The ester is isolated as in Method *H*.

## 7. Esters of Anthranilic Acid

### *Method J (Staiger and Miller*[42]*)*

*Lower alcohols.* Prepare a mixture of isatoic anhydride/alcohol/sodium hydroxide (molar ratio 1/3/0·05). Stir the mixture and raise the temp slowly to approx. 65°C, at which temp moderate evolution of carbon dioxide occurs. Maintain the temp until the anhydride has been consumed and the evolution of carbon dioxide has ceased. Cool the mixture, and dilute with three volumes of water. The ester settles out as an oil which is separated and converted into the picrate.

*Higher alcohols.* Dissolve the reactants in the molar ratio 1/1/0·05 (for dihydric alcohols 2/1/0·05) in dioxan, stir the mixture and raise the temp slowly until a moderate evolution of carbon dioxide occurs, usually between 60°C and 100°C, then continue as above. Solid esters are purified by recrystallization from methanol or aqueous methanol.

For reaction scheme see p. 33.

## 8. Esters of Pyruvic Acid 2,6-Dinitrophenylhydrazone

### *Method K (Schwartz and Brewington*[51]*)*

Dissolve 10 g of chloro-2,6-dinitrobenzene in 50 ml of 50% alcohol. Add an equivalent amount of hydrazine hydrate in small portions through a reflux condenser, so that the reaction does not become too vigorous. Filter the hot solution at this point to remove traces of a black impurity, reflux for 10–15 min and allow to cool slowly to room temp. When the dark red 2,6-dinitrophenylhydrazine is accompanied by an orange red impurity, the latter is removed by washing on the filter with hot water. Yield about 50%, m.p. 143–144°C.

Dissolve 5 g 2,6-dinitrophenylhydrazine in 30 ml water at 80°C. Add an equimolar amount of pyruvic acid (2·2 g) and stir mechanically for 1 h, cool slowly to room temp and filter the bulky yellow precipitate off, wash with cold (4°C) water and dry thoroughly. Suspend 5 g of the dry product in 10 ml thionyl chloride, reflux under magnetic stirring until clear, cool slowly to room temp, allow to stand for 1 h at 4°C, filter with suction and dry in vacuum at room temp until odourless. M.p. 156–158°C.

Prepare a stock solution of the acid chloride by dissolving 2·35 g (8 m-mole) in 100 ml purified benzene. Pipet 10 ml into a dry flask and add the alcohol (4–5 drops for alcohols containing 7 carbon atoms or less, 0·5 m-mole for higher alcohols), add 5 ml of a benzene solution of triethylenediamine (containing 19·4 mg per ml of purified benzene) in 1 ml portions, swirling the flask briefly between the additions, and allow the mixture to stand at room temp for 15–20 min. Prepare in the meantime a column of alumina by adding 15 g of 8% acidic alumina to a chromatography tube (1·8 × 13 cm), half-filled with benzene. Drain the excess benzene, and when just dry pour carefully the reaction mixture onto the column and start collecting the effluent. Develop by adding benzene until all the colour is removed below the band of excess reagent which remains at the top of the column. Evaporate the effluent on the steam bath under a gentle stream of nitrogen and recrystallize from absolute ethanol.

## 9. Aryl 2,4-Dinitrophenyl Ethers

### *Method L (Bost and Nicholson*[56]*)*

Dissolve 0·01 mole of the phenol in 5 ml 2 N sodium hydroxide. Add 0·01 mole (2 g) of chloro-2,4-dinitrobenzene dissolved in 30 ml ethanol. Reflux the intensely coloured solution on the steam-bath until the colour fades and a precipitate has formed (about 30 min). Add an equal volume of water, cool, isolate the precipitate by filtration, wash with water and recrystallize from ethanol.

### *Method M (Chapman and Parker*[58]*)*

Dissolve equivalent quantities (about 3 m-mole) of the phenol and of fluoro-2,4-dinitrobenzene in 5 ml benzene. Add 0·1–0·5 ml triethylamine as catalyst, and leave the mixture to stand overnight at room temp or reflux on the steam-bath for 1–4 h. Cool and isolate the ether by filtration. Purify by recrystallization from ethanol or aqueous ethanol.

### *Method N (Reinheimer et al.*[59]*)*

Dissolve 5 m-mole of the phenol in 5 ml acetone. Add 0·5 ml triethylamine and 0·93 g of fluoro-2,4-dinitrobenzene in 5 ml acetone. Reflux the mixture on the water-bath for 30 min, evaporate the acetone, add 10 ml of 5% hydrochloric acid, isolate the precipitate by filtration, and wash with water. Heat the residue with 15 ml 5% sodium hydroxide, filter off and wash with water. Dry the residue for 24 h *in vacuo* over magnesium sulphate and recrystallize from ethanol, aqueous ethanol, or glacial acetic acid.

### *Method O (Vorozhtsov, Jr. and Iacobson*[60]*)*

Mix 2·5 m-mole of the phenol with 0·47 g fluoro-2,4-dinitrobenzene (or 0·51 g chloro-2,4-dinitrobenzene) and 0·29 g (0·05 mole) anhydrous potassium fluoride and heat the mixture for some time at 100–110°C. Extract the reaction mixture with hot benzene, filter, evaporate the benzene solution on the steam-bath and recrystallize the residue from ethanol.

## 10. Alkyl Trimethylsilyl Ethers

### *Method P (Langer et al.*[62]*)*

Reactive alcohols. To 0·2 mole of the alcohol in a flask (equipped with reflux condenser and drying tube) add 0·11 mole of hexamethyldisilazane. Raise the temp to 93°C over a period of 5 h, during which ammonia is evolved. Distil the residue and redistil the crude product through a Podbielniak column.

*Less reactive alcohols.* Add to a mixture of 0·18 mole of the alcohol 0·1 mole of hexamethyldisilazane and two drops of chlorotrimethylsilane. A precipitate is formed immediately. Raising the temp, evolution of ammonia starts at 60°C. After 2 h a temp of 132°C is obtained, at which time the precipitate is seen to sublime out of the flask. After 4 h, the reflux temp is 204°C. Cool and purify as above.

For diols the mole ratio is 0·15/0·17, and 3 drops of chlorotrimethylsilane are added.

*Phenols.* Reflux 0·4 mole of the phenol and 0·5 mole of chlorotrimethylsilane for 7 h, cool, add 50 ml petroleum ether b.p. 60–68°C, reflux for 4 h, then proceed as indicated above for reactive alcohols.

According to Wildenhain *et al.*[(63)] an addition of formamide facilitates the silylation of phenols.

### *Method Q (Freedman and Croitoru*[(67)]*)*

For hindered phenols acidic oxides are the best catalysts for silylation. Place 1 g (or 1 ml) of the sample, 1·5 g hexamethyldisilazane, 0·5 g $TiO_2$ (or $Al_2O_3$) and, if desired, 0·5 g drierite in a flask fitted with an air condenser. Heat the flask for 45 min on a 200°C sand-bath or heating mantle. After cooling, the residue is purified as above or a sample is injected immediately into a gas chromatograph. (See below, Part C, Chapter 10.13 *c* (iv) 2, p. 108 and recommended Method *R* (*b*), p. 141.)

### *Method R (Friedman and Kaufman*[(69)]*)*

Silylation of tertiary alcohols. Reflux a mixture of 0·1 mole of the alcohol, 0·11 mole hexamethyldisilazane and 50 ml of freshly distilled dimethylformamide overnight. Add a small amount of tridecanol and reflux for 2 h to destroy excess of hexamethyldisilazane, if any. Then distil the solution through a 40-cm vacuum-jacketed packed column.

Another procedure is to reflux 0·2 mole of the alcohol, 0·22 mole hexamethyldisilazane and 50 ml dimethylsulphoxide overnight. During reflux two phases separate, the upper phase being essentially pure alkoxytrimethylsilane. This phase is purified by distillation.

## 11. Aryloxyacetic Acids

### *Method S (Koelsch*[(70)]*)*

Add 1 g of the phenol to 4 ml 10 N sodium hydroxide in a test-tube or a 25-ml Erlenmeyer flask, followed by 1·25 g monochloroacetic acid, 1–2 ml water being added if solution does not occur. Heat the mixture on the water-bath for 1 h, cool, acidify with dilute sulphuric acid (indicator methyl orange) and extract the solution with 3 × 5 ml portions of ether. Wash the combined ether extracts once with water, then with 10 ml 2 N sodium carbonate solution. Acidify the sodium carbonate layer and isolate the aryloxyacetic acid by filtration. If precipitation does not occur add some solid sodium chloride, which may provoke precipitation. Purify by recrystallization from water.

## 12. S-Alkylthiouronium Picrates

### *Method T (Schotte and Veibel*[(74)]*)*

Convert the tertiary alcohol into alkyl halide as indicated above (p. 9). Dissolve 1·52 g thiourea in 4 ml water, add 3 ml ethanol and 2 m-mole (1–2 ml) alkyl halide and reflux the mixture on the steam-bath until the alkyl halide has disappeared, then for 15 min further (usually 2–3 h). Pour the solution into 200–300 ml of 1% aqueous solution of picric acid (or a solution of 3,5-dinitrobenzoic acid, Jureček *et al.*[(75)]); filter off the precipitate of S-alkylthiouronium picrate (or of 3,5-dinitrobenzoate) and recrystallize from ethanol or aqueous ethanol.

# REFERENCES TO PART B


1. GOLDSCHMIDT, H. and MEISSLER, A., *Ber.* **23**, 257 (1890).
2. WILLSTÄTTER, R. and HOCHEDER, F., *Liebigs Annalen* **354**, 253 (1907).
3. JOHNSON, J. D. A., *J. Chem. Soc.* **1955**, 3322.
4. HOEKE, F., *Rec. Trav. chim. Pays-Bas* **54**, 505 (1935).
5. VEIBEL, S. and LILLELUND, H., *Dansk Tidsskr. Farm.* **14**, 236 (1940); *Chem. Abs.* **35**, 2444 (1941).
6. VEIBEL, S., LILLELUND, H. and WANGEL, J., *Dansk Tidsskr. Farm.* **17**, 183 (1943); *Chem. Abs.* **39**, 1608 (1945).
7. REKKER, R. F., FABER, A. C., TOM, D. H. E., VERLEUR, H. and NAUTA, W. T., *Rec. Trav. chim. Pays-Bas* **70**, 113 (1951).
8. FRITZ, J. S. and SCHENK, G. H., *Anal. Chem.* **31**, 1808 (1959).
9. SPASSOW, A., *Ber.* **70**, 1926 (1937); **75**, 779 (1942).
10. SCHOTTEN, C., *Ber.* **17**, 2544 (1884).
11. BAUMANN, E., *Ber.* **19**, 3218 (1886).
12. HENSTOCK, H., *J. Chem. Soc.* 216 (1933).
13. KOMÁREK, K., CHURÁČEK, J. and VANASEK, V., *Sb. Ved. Pr. Vys. Sk. Chemickotechnol. Pardubice* **21**, 71 (1969); *Chem. Abs.* **73**, ref. 94386 (1970).
14. REICHSTEIN, T., *Helv. Chim. Acta* **9**, 799 (1926).
15. SAUNDERS, B. C., STACEY, G. J. and WILDING, I. G. E., *Biochem. J.* **36**, 368 (1942).
16. DUNBAR, R. E. and FERRIN, F. J., *Chemist-Analyst* **44**, 77 (1955).
17. HUTH, S. D., *May and Baker Laboratory Bulletin* **2**, 53 (1957).
18. GARSKA, K. J., DOUTHIT, R. C. and YARBOROUGH, V. A., *Anal. Chem.* **33**, 392 (1961).
19. JENSEN, K. A., LIMBORG, F. and SOLSTAD, W. M., *Acta Chem. Scand.* **4**, 392 (1950).
20. VAN DUIN, H., *Rec. Trav. Chim. Pays-Bas* **73**, 68 (1954).
21. FREYTAG, W. and BAUSTIAN, M., *J. Chromatog.* **13**, 558 (1964).
22. POHLOUDEK-FABINI, R. and BEYRICH, T., *Pharmazie* **15**, 356 (1960).
23. BEYRICH, T. and POHLOUDEK-FABINI, R., *Pharmazie* **15**, 437 (1960).
24. POHLOUDEK-FABINI, R. and MÜNCHOW, P., *Pharmazie* **19**, 591 (1964).
25. HECKER, E., *Chem. Ber.* **88**, 1666 (1955).
26. BOSVIK, R., KNUTSEN, K. V. and VON SYDOW, C. F. E., *Anal. Chem.* **33**, 1162 (1961).
27. AMIN, EL. S. and SAMAD, M. ABD. EL., *Mikrochim. Acta* 374 (1969).
28. VEIBEL, S. and PEDERSEN, C., *Acta Chem. Scand.* **9**, 1674 (1955).
29. NARANG, C. K. and MATHUR, N. K., *Indian J. Chem.* **3**, 182 (1965); *Chem. Abs.* **63**, 10687 (1965); *Indian J. Chem.* **4**, 263 (1966); *Chem. Abs.* **65**, 12852 (1966).
30. FESSLER, W. A. and SHRINER, R. L., *J. Am. Chem. Soc.* **58**, 1384 (1936).
31. LAWLOR, F. E., *J. Ind. Eng. Chem.* **39**, 1419 (1947).
32. ELVING, P. J. and WARSHOWSKY, B., *Anal. Chem.* **19**, 1006 (1947).
33. DICKINSON, G. M., CROSSON, L. H. and COPENHAVER, J. E., *J. Am. Chem. Soc.* **59**, 1094 (1937).
34. DE GRAEF, H. and PIERRET, J., *Bull. Soc. Chim. Belg.* **57**, 307 (1948).

35. Floria, J. A., Dobratz, I. W. and McClure, J. M., *Anal. Chem.* **36**, 2053 (1954).
36. Siggia, S., Hanna, J. G. and Culmo, R., *Anal. Chem.* **33**, 900 (1961).
37. Siggia, S., *Quantitative Organic Analysis via Functional Groups,* 3rd Edn., p. 26, Wiley & Sons, New York (1963).
38. Harper, R., Siggia, S. and Hanna, J. G., *Anal. Chem.* **37**, 600 (1965).
39. Salmon-Legagneur, F. and Neveu, C., *Bull. Soc. Chim. France* 1309 (1963).
40. Salmon-Legagneur, F., *Bull. Soc. Chim. France* 580 (1952).
41. Oppenauer, R. V., *Monatsh. Chem.* **97**, 67 (1966).
42. Staiger, R. P. and Miller, E. B., *J. Org. Chem.* **24**, 1214 (1959).
43. Teeter, R. M., *Anal. Chem.* **38**, 1736 (1966).
44. Mesnard, P., Gibirila, B. and Bertucat, M., *Chim. Anal.* (*Paris*) **45**, 491 (1963).
45. Mesnard, P., Gibirila, B. and Bertucat, M., *Kongr. Pharm. Wiss. Vortr. Original-mitt.* **23**, 603 (Münster, 1963).
46. Kharasch, N., McQuarrie, D. P. and Buess, C. M., *J. Am. Chem. Soc.* **75**, 2658 (1953).
47. Lefèbvre, G., Berthelin, J., Maugras, M., Gay, R. and Urion, E., *Bull. Soc. Chim. France* **1966**, 266.
48. Iyer, V. and Mathur, N. K., *Anal. Chim. Acta* **33**, 554 (1965).
49. Narang, C. K., Iyer, V. and Mathur, N. K., *Microchem. J.* **9**, 408 (1965).
50. Berger, J. and Uldall, I., *Acta Chem. Scand.* **18**, 1353 (1964).
51. Schwartz, D. P. and Brewington, C. R., *Microchem. J.* **11**, 430 (1966).
52. Wekell, J. C., Houle, C. R. and Malins, D. C., *J. Chromatog.* **14**, 529 (1964).
53. Borecký, J., *Chem. & Ind.* (*London*) 265 (1962).
54. Borecký, J., *Collection Czech. Chem. Commun.* **27**, 2761 (1962).
55. Mathur, N. K., *Talanta* **13**, 1601 (1966).
56. Bost, R. W. and Nicholson, F., *J. Am. Chem. Soc.* **57**, 2368 (1935).
57. Whalley, W. B., *J. Chem. Soc.* 2241 (1950).
58. Chapman, N. B. and Parker, R. E., *J. Chem. Soc.* 3301 (1951).
59. Reinheimer, J. D., Douglass, J. P., Leister, H. and Voelkel, M. B., *J. Org. Chem.* **22**, 1743 (1957).
60. Vorozhtsov, Jr., N. N. and Iacobson, G. G., *Zh. obshch. Khim.* **28**, 40 (1958); *J. Gen. Che.* (*USSR*) **28**, 40 (1958).
61. Cohen, I. C., Norcup, J., Ruzicka, J. H. A. and Wheals, B. B., *J. Chromatog.* **44**, 251 (1969).
62. Langer, S. H., Connell, S. and Wender, I., *J. Org. Chem.* **23**, 50 (1958).
63. Wildenhain, W., Heinichen, G. and Henseke, G., *J. Prakt. Chem.* **20**, 35 (1963).
64. Wood, R., *J. Gas. Chromatog.* **6**, 94 (1968).
65. Leibman, K. C. and Ortiz, E., *J. Chromatog.* **32**, 757 (1968).
66. Langer, S. H., Pantages, P. and Wender, I., *Chem. & Ind.* (*London*) 1661 (1958).
67. Freedman, R. W. and Croitoru, P. P., *Anal. Chem.* **36**, 1389 (1964).
68. Blakley, E. R., *Anal. Biochem.* **15**, 350 (1966).
69. Friedman, S. and Kaufman, M. L., *Anal. Chem.* **38**, 144 (1966).
70. Koelsch, C. F., *J. Amer. Chem. Soc.* **53**, 304 (1931).
71. Meadoe, J. R. and Reid, E. E., *J. Am. Chem. Soc.* **65**, 457 (1943).
72. Böhme, H. and Opfer, H., *Z. anal. Chem.* **139**, 255 (1953).
73. Jesurun, J. A., *Ber.* **26**, 2286 (1893).
74. Schotte, L. and Veibel, S., *Acta. Chem. Scand.* **7**, 1357 (1953).
75. Jureček, M., Chládek, O., Chládková, R., Souček, M., and Srpová, B., *Collection Czech. Chem. Commun.* **22**, 1809 (1957).
76. Prajsnar, B., *Chem. Anal.* (*Warsaw*) **6**, 1039 (1961); *Chem. Abs.* **57**, 15 (1962).
77. Ritter J. J. and Minieri P. P. *J. Am. Chem. Soc.* **70**, 4045 (1948).

78. DUKE, F. R. and WITMAN, R. C., *Anal. Chem.* **20**, 490 (1948).
79. SAMUELSON, O. and STRÖMBERG, H., *Acta Chem. Scand.* **22**, 1252 (1968).
80. MUKAI, K., HASHIMOTO, A. and YODA, N., *Yukagaku* **12**, 505 (1963); *Chem. Abs.* **60**, 6195 (1964).
81. SHERMA, J. and THOMPSON, JR., D. E., *Anal. Chim. Acta* **32**, 181 (1965).
82. BROWNELL, H. H., *J. Chromatog.* **19**, 556 (1965).
83. BORECKÝ, J. and GASPARIČ, J., *Collection Czech. Chem. Commun.* **25**, 1287 (1960).
84. APRIL, A., *Ingr. Chimiste* **43**, 209 (1961); *Chem. Abs.* **62**, 11155 (1965).
85. MACÁK, J. and KUBEŠ, J., *Collection Czech. Chem. Commun.* **25**, 301 (1960).
86. GASPARIČ, J., PETRÁNEK, J. and BORECKÝ, J., *J. Chromatog.* **5**, 408 (1961).
87. LOCKE, D. and SHERMA, J., *Anal. Chim. Acta* **25**, 312 (1961).
88. CLARK, I. T., *J. Chromatog.* **15**, 65 (1964).
89. FUJIWARA, F. and SUGIMOTO, C., *Kagaku to Kôgyô* **32**, 265 (1958); *Chem. Abs.* **53**, 4012 (1959).
90. AARNA, A. and KARJAMA, E., *Tr. Tallin Politekh. Inst. Ser. A*, **254**, 13 (1967); *Chem. Abs.* **70**, ref. 92996 (1969).
91. MICHEEL, F. and LEIFELS, W., *Mikrochim. Acta* 467 (1961).
92. MICHEEL, F. and SCHMINKE, W., *Chem. Ber.* **91**, 984 (1958).
93. GUMPRECHT, D. L., *J. Chromatog.* **30**, 528 (1967); **37**, 268 (1968).
94. POHLOUDEK-FABINI, R. and BEYRICH, T., *Naturwissenschaften* **44**, 512 (1957).
95. POHLOUDEK-FABINI, R. and BEYRICH, T., *Pharm. Zentralhalle* **99**, 341 (1960).
96. GASPARIČ, J. and BORECKÝ, J., *J. Chromatog.* **4**, 138 (1960).
97. POZETTI, G. L., *Rev. Fac. Farm. Odontol. Araraquara* **2**, 223 (1968); *Chem. Abs.* **72**, ref. 62542 (1970).
98. BERBALK, H., *Monatsh. Chem.* **90**, 24 (1959).
99. FRAHN, J. L., *J. Chromatog.* **37**, 279 (1968).
100. ATTAWAY, J. A., WOLFORD, R. W., ALBERDING, G. E. and EDWARDS, G. J., *Anal. Chem.* **34**, 671 (1962).
101. ATTAWAY, J. A., WOLFORD, R. W., ALBERDING, G. E. and EDWARDS, G. J., *Anal. Chem.* **35**, 234 (1963).
102. MICHEEL, F. and SCHMINKE, W., *Angew. Chem.* **69**, 334 (1957).
103. GASPARIČ, J. and BORECKÝ, J., *J. Chromatog.* **5**, 466 (1961).
104. GAUTHIER, H. and MANGENEY, G., *J. Chromatog.* **14**, 209 (1964).
105. CHURÁČEK, J., *J. Chromatog.* **48**, 241 (1970).
106. JACKSON, M. and HAYWARD, L. D., *J. Chromatog.* **5**, 166 (1961).
107. TULUS, M. R. and GÜRAN, A., *Arch. Pharm.* **296**, 623 (1963).
108. MARUTA, S., SUZUKI, Y. and IWAMA, F., *Kogyo Kagaku Zasshi* **63**, 2140 (1960); *Chem. Abs.* **57**, 1554 (1962).
109. CRUMP, G. B., *J. Chromatog.* **10**, 21 (1963).
110. MAKES, J., *Chem. Prum.* **18**, 30 (1968); *Chem. Abs.* **68**, ref. 92775 (1968).
111. CHURÁČEK, J., KOMÁREK, K., VAŇÁSEK, V. and JUREČEK, M., *Collection Czech. Chem. Commun.* **33**, 3876 (1968).
112. SUCKER, H., *Deut. Apotheker-Ztg.* **104**, 947 (1964).
113. STONE, H. and SHECHTER, H., *J. Org. Chem.* **15**, 491 (1950).
114. SUBBARAO, R., ROOMI, M. W., SUBBARAM, M. R. and ACHAYA, K. T., *J. Chromatog.* **9**, 295 (1962).
115. PURDY, S. J. and TRUTER, E. V., *J. Chromatog.* **14**, 62 (1964).
116. NOBUHARA, A., *J. Chromatog.* **30**, 235 (1967).
117. BANDYOPADHYAY, C. and CHAKRABARTY, M. M., *J. Chromatog.* **32**, 297 (1968).
118. MCSWEENEY, G. P., *J. Chromatog.* **17**, 183 (1965).
119. STAHL, E. and VOLLMANN, H., *Talanta* **12**, 525 (1965).

120. KNAPPE, E., PETERI, D. and ROHDEWALD, I., *Z. anal. Chem.* **199**, 270 (1964).
121. FISCHER, F. and KOCH, H., *J. Chromatog.* **16**, 246 (1964).
122. DE SIMONE, V. and VICEDOMINI, M., *J. Chromatog.* **37**, 538 (1968).
123. AKHREM, A. A., KUZNETSOVA, A. I., TITOV, YU. A. and LEVINA, I. S., *Izv. Akad. Nauk SSSR, Otd. Khim. Nauk* **1962**, 657; *Chem. Abs.* **57**, 4003 (1962).
124. LIPINA, T. G., *Tr. po. Khim. i Khim. Tekhnol.* 424 (1962); *Chem. Abs.* **59**, 8100 (1963).
125. VAZQUEZ, A. T. and MONTES, A. L., *Anales Asoc. Quim. Arg.* **51**, 231 (1963); *Chem. Abs.* **63**, 9739 (1965).
126. TADEMA, G. and BATELAAN, P. H., *J. Chromatog.* **33**, 460 (1968).
127. DITTMANN, J., *J. Chromatog.* **32**, 764 (1968).
128. CHAFETZ, L., KAY, A. I. and SCHRIFTMAN, H., *J. Chromatog.* **35**, 567 (1968).
129. VAN SUMERE, C. F., WOLF, G., TEUCHY, H. and KINT, J., *J. Chromatog.* **20**, 48 (1965).
130. ALY, O. M., *Z. anal. Chem.* **234**, 251 (1968).
131. BARK, L. S. and GRAHAM, R. J. T., *J. Chromatog.* **27**, 109, 116, 131 (1967).
132. GRAHAM, R. J. T., *J. Chromatog.* **33**, 118 (1968).
133. DIAMOND, P. F., *J. Chromatog.* **32**, 419 (1968).
134. CRABTREE, A. N. and MCGILL, A. E. J., *Mikrochim. Acta* **1967**, 85.
135. PARIHAR, D. B., SHARMA, S. P. and TEWARI, K. C., *J. Chromatog.* **24**, 230 (1966).
136. VAN DIJK, J. H. and MIJS, W. J., *Z. anal. Chem.* **236**, 419 (1968).
137. RÖHR, M. and CHIARI, D., *Mikrochim. Acta* 137 (1967).
138. DERBESY, M. and NAUDET, M., *Bull. Soc. Chim. France* 2962 (1968).
139. DALLAS, M. S. J., *J. Chromatog.* **48**, 225 (1970).
140. DHONT, J. H. and DE ROOY, C., *Analyst* **86**, 527 (1961).
141. DHONT, J. H. and DIJKMAN, G. J. C., *Chem. Techn. Rev.* **22**, 521 (1967); *Chem. Abs.* **69**, ref. 35298 (1968).
142. MEHLITZ, A., GIERSCHNER, K. and MINAS, T., *Chemiker-Ztg.* **89**, 175 (1965); *Chem. Abs.* **63**, 1154 (1965).
143. SEVERIN, M., *J. Chromatog.* **26**, 101 (1967).
144. MALINS, D. C., WEKELL, J. C. and HOULE, C. R., *Anal. Chem.* **36**, 658 (1964).
145. BUZLANOVA, M. M., UL'YANOVA, V. N. and OBTEMPERANSKAYA, S. I., *Zh. Anal. Khim.* **23**, 1425 (1968); *Chem. Abs.* **69**, ref. 113355 (1968).
146. WILDENHAIN, W. and HENSEKE, G., *J. Chromatog.* **19**, 438 (1965).
147. PARIHAR, D. B., SHARMA, S. P. and TEWARI, K. C., *J. Chromatog.* **21**, 261 (1966).
148. LEIBMAN, K. C. and ORTIZ, E., *J. Chromatog.* **32**, 757 (1968).

# Part C

# Estimation of Hydroxy Compounds

## CHAPTER 9

# DIRECT TITRATION OF HYDROXY COMPOUNDS AS ACIDS

ALIPHATIC hydroxy compounds are usually too weakly acidic to allow their direct titration, even in non-aqueous solution. In contrast, aromatic hydroxy compounds can be titrated, usually potentiometrically but in some instances also by visual titration, when a suitable solvent and titrant are chosen.

In a study of potentiometric titrations in non-aqueous solvents, van der Heijde and Dahmen[(1)] found that a potential jump of 200–300 mV between the half-neutralization potential and the next plateau is sufficient to allow satisfactory accuracy and they suggested that the following factors should be considered before making a choice of solvent:

1. The acidity or basicity of the solvent (its levelling effect).
2. The intrinsic acidity of the acid to be titrated.
3. The tendency of the ions involved to form particularly stable complexes or insoluble salts.
4. Liquid junction potentials at interfaces.
5. The dielectric constant of the solvent.

In subsequent papers, van der Heijde[(2, 3)] pointed out that anomalies may arise from the association of the anions of an acid or their adsorption on the surface of the electrode, and he recommended[(3)] the following solvents: glacial acetic acid, acetone, acetonitrile, propan-2-ol, water, dimethylformamide, pyridine, ethylenediamine.

Figure 2, taken from the paper of van der Heijde and Dahmen,[(1)] shows the potential range covered by each of these solvents.

Streuli[(4)] found that very weak phenols have steeper titration curves in pyridine than in water. Crabb and Critchfield[(5)] compared solvents for differentiating titrations of phenols and found the following series (in order of decreasing differentiating power): *tert*-butanol > pyridine > tetra-

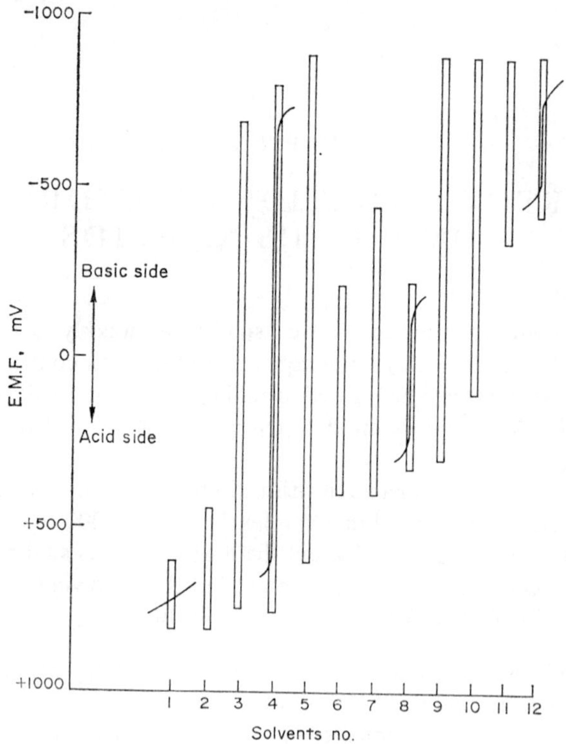


FIG. 2. Approximate potential ranges of twelve solvents arranged in the order of increasing basicity; with examples of titration curves schematically indicated.

Solvents

| | |
|---|---|
| 1. Trifluoroacetic acid | 7. Propan-2-ol |
| 2. Acetic acid | 8. Water |
| 3. Chlorobenzene | 9. Dimethylformamide |
| 4. Acetone | 10. Pyridine |
| 5. Acetonitrile | 11. Butylamine |
| 6. Methanol | 12. Ethylenediamine |

(van der Heijde and Dahmen[1])

(Reproduced by kind permission of the authors and Elsevier Publishing Co.)

hydrofuran > propan-2-ol > dimethylsulphoxide > acetone > acetonitrile > ethylene glycol > water.

For visual titration of phenols Fritz and Keen[6] recommended dimethylformamide as solvent and azoviolet as indicator for negatively substituted phenols and ethylenediamine, with *o*-nitroaniline as indicator, for phenol and alkyl-substituted phenols, the titrants in both instances being potassium methoxide in methanol/benzene (see recommended Method *A*, p. 113).

Coetzee and Padmanabhan(7) pointed out that the titration curves of solutions of phenols in acetonitrile show some deformation due to the formation of homoconjugated complexes $BH \cdots B^-$ and $(BH)_2B^-$ (cf. van der Heijde(2)). A complication linked with the use of acetonitrile as solvent is that the strongest base to be used is the phenoxide ion, which shows a strong tendency to homoconjugation. Attempts to use stronger bases, e.g. $^-CH_2CN$, result in polymerization of the solvent.

As solvent for the potentiometric titration (glass electrode/calomel electrode assembly) of very weak acids and bases, Morman and Harlow(8) recommended sulpholan, 3-methylsulpholan, or 2,4-dimethylsulpholan as these solvents have an adequate dielectric constant (about 44), little tendency to hydrogen bonding, and very weak acidic and basic properties. Even strongly sterically hindered phenols such as 2,6-di-*tert*.-butyl-4-methylphenol can be titrated. The potential span from strong acids to strong bases is unusually great (from −700 to +1000 mV). 3-Methylsulpholan is the preferred solvent, as it is liquid at room temperature and more easily purified than 2,4-dimethylsulpholan. The procedure is described below, recommended Method *B*, p. 114.

Harlow(9) has found some anomalies in the titration curve near the end-point when very weak acids are titrated. The cause of these anomalies has been traced to the presence of potassium or sodium ions, which react with the tip of the glass electrode; quaternary ammonium bases should therefore be used as titrants instead of potassium or sodium methoxide.

For the preparation of tetrabutylammonium hydroxide two methods have been recommended.

### *Method 1. Sensabaugh, Cundiff and Markunas*(10)

Dissolve 80 g tetrabutylammonium iodide in 180 ml anhydrous methanol, cool the solution in an ice-bath, stir mechanically, and add 40 g finely powdered silver oxide, the stirring being continued for 1 h. Filter the suspension through a fine-pored sintered-glass funnel, wash the funnel with 3 × 50-ml portions of benzene and make the combined filtrate and washings up to 2000 ml with benzene.

For purification, prepare a column of Amberlite IRH-400 ($OH^-$).

Half-fill a chromatography tube, 25 × 400 mm, with Amberlite IRH-400. Wash the resin, first with 2 N sodium hydroxide until the washings from the column are free from halogen ions, then with water until the washings are neutral. Pour slowly 500 ml methanol onto the column, followed by 500 ml benzene/methanol (10/1). Then pour the tetrabutylammonium hydroxide solution onto the column at a rate such that 15–20 ml will wash off the column per min. The washings are collected in a flask, protected from carbon dioxide and moisture, as soon as they become alkaline.

### *Method 2. Harlow and Wyld*[11], *Harlow and Bruss*[12]

To 100 ml of 0·2 N potassium hydroxide in propan-2-ol, add 10–20% excess of dry quaternary ammonium chloride. Shake the flask vigorously at intervals over a period of 15 min, the allow the precipitate of potassium chloride to settle. Decant the supernatant solution into another flask, preferably one which permits delivery of the titrant yet protects it from exposure to carbon dioxide.

A disadvantage of this method is that it yields titrants with a relatively high content of potassium ions, 100–200 mg of potassium chloride per litre, and according to Harlow[9] this may cause serious difficulties in the titration of very weak acids (see p. 71). The content of potassium ions may, however, be reduced to negligible concentrations by treating the titrant with a strong acid ion exchange resin in its quaternary ammonium ion form.

Standardize the titrant against benzoic acid dissolved in dimethylformamide, either potentiometrically or with thymol blue indicator. As dimethylformamide always contains acidic impurities, including carbon dioxide, the solvent must either be neutralized to the thymol blue end-point before use or a blank with the same amount of solvent carried out.

Buell[13] used this procedure for differential titration of phenols in petroleum. Dissolve about 0·01 m-equiv. of phenols in 10 ml benzene, add 10 ml pyridine and titrate with 0·03 N tetrabutylammonium hydroxide. Near the end-points add the titrant in 0·02 ml increments.

Diembeck *et al.*[14] used 0·1 N tetramethylammonium hydroxide in *tert.*-butanol/ethanol for the titration of monohydric phenols with p$K_a$-values (in water) between 0·71 and 10·58, dissolved in non-aqueous solvents (acetone, acetonitrile, *tert.*-butanol, dimethylformamide, dimethylsulphoxide, methyl ethyl ketone, isobutyl methyl ketone, pyridine). The titration was carried out potentiometrically, conductometrically or visually with an accuracy of within ±1%.

Fritz and Gainer[15] studied thirteen indicators for visual titration of weak acids (including phenols) in pyridine solution with tetrabutylammonium hydroxide prepared according to Harlow *et al.*[16] and found thymol blue, 2,4-dinitrodiphenylamine, 4-nitro-4′-aminoazobenzene, *o*-nitroaniline and azoviolet satisfactory. They also studied[15] coulometric titration of acids (including phenols) in *tert.*-butyl alcohol or in methyl alcohol with electrolytically generated tetrabutylammonium hydroxide, either potentiometrically or visually. Amounts of acids ranging from 10 to 60 microequivalents were determined with satisfactory precision and accuracy.

Belcher *et al.*[17] used 0·01 N solution of tetrabutylammonium hydroxide in benzene/methanol for submicro determination of phenols in pyridine solution, visually or potentiometrically with a glass/silver electrode assembly.

*Visual titration.* Weigh the sample (30–80 μg) into the titration vessel and dissolve it in 0·3 ml pyridine. Add 0·02–0·03 ml thymol blue indicator (0·08 g in 50 ml propan-2-ol) and titrate to the first appearance of a blue colour. Direct a slow stream of nitrogen over the solution during the titration, and stir magnetically. Carry out a blank titration in the same way.

*Potentiometric titration.* Weigh the sample (30–80 μg) into the titration vessel and dissolve it in 0·3 ml pyridine. Add a glass-encased rotor for magnetic stirring, adjust the glass electrode so that its tip is 1–2 mm below the liquid surface, and place the burette and the capillary for nitrogen introduction in position. Add a little thymol blue solution to give warning of the appearance of the end-point. Titrate with 0·01 N tetrabutylammonium hydroxide by adding 5 μl portions and allowing the potential to reach equilibrium between additions. Near the end-point add 1·0 μl portions of the titrant.

Aksenenko *et al.*(18) found that β-nitrosubstituted alcohols are sufficiently acidic to permit their estimation by titration. Dissolve 0·1 g of the alcohol in 40 ml acetone and titrate with 0·1 N tetrabutylammonium hydroxide in methanol/dioxan or with 0·1 N tetraethyl or tetramethylammonium hydroxide in benzene/methanol, potentiometrically or visually. As indicator use 1% ethanolic solution of crystal violet or thymolphthalein.

For preparation of the tetrabutylammonium hydroxide solution dissolve under nitrogen 100 g of tetrabutylammonium hydroxide in 300 ml methanol, filter the solution, and add 700 ml dioxan to the filtrate. Leave the solution to stand for 1–2 days and then standardize it against benzoic acid.

Fialko *et al.*(18a) have recently suggested the use of tetrabutylammonium hydroxide as titrant and a mixture of benzene and dimethylsulphoxide as solvent for the potentiometric titration of alkylphenols. The titration is carried out in an atmosphere of dry nitrogen, using a glass electrode/calomel electrode assembly, the calomel electrode containing saturated potassium chloride in propan-2-ol. The potential jump is distinct only in a benzene/dimethylsulphoxide (1/9) mixture.

In polyhydric phenols, the number of hydroxyl groups titrated depends on the structure of the molecule. One hydroxyl group is titrated in dihydric phenols and in compounds where the alkylhydroxyphenyl radicals are linked with sulphide or methylene bridges in the position *ortho* to the hydroxyl group. When the hydroxyl group is in the *p*-position with respect to the methylene bridge, then all the hydroxyl groups are titrated quantitatively.

Greenhower and Smith(19) have given the following method for the determination of phenols in coal-tar:

Dissolve 0·1 g coal-tar in 20 ml pyridine and titrate potentiometrically with 0·2 N sodium β-aminoethanolate in ethylenediamine, using an antimony indicating electrode and a platinum reference electrode. Add the titrant in 0·1 ml increments and allow the potential to reach equilibrium between increments (2–15 min). Each titration takes about 2 h to complete.

Pandya and Haldar[20] as well as Vaughan and Swithenbank[21] have studied thermometric titration of phenols with non-aqueous solutions of alkali. A drawback is that the titration curves often show only a slight change in slope at the end-point. Vaughan and Swithenbank[21] found that if acetone is used as solvent for the acid it will act as an indicator because at the neutralization point a rapid development of heat occurs through the formation of diacetone alcohol. Even sterically hindered phenols can be estimated. Absence of water is essential; 0·2% water decreases the accuracy considerably, and 2% water invalidates the estimation.

Vaughan and Swithenbank[22] used this method for the estimation of phenolic hydroxyl groups in coal. The procedure is given below (recommended Method *C*, p. 114).

The direct titration of aliphatic hydroxy groups that are not particularly activated by electronegative neighbouring groups demands not only non-aqueous solvents but also special titrants. Fritz[23] studied the titration of enols dissolved in dimethylformamide with sodium methoxide in benzene/methanol, either visually with azoviolet indicator, or potentiometrically. He found that compounds A—$CH_2$—A′ can be titrated when A and A′ are —COR, —CHO, —COOR, —CONHR or —CN, whereas the —$CONH_2$ group is not sufficiently electron-attracting to permit the formation of an enol (or the removal of a hydrogen ion from the $CH_2$-group). Cyanoacetamide cannot be titrated by this procedure, but if it is dissolved in ethylenediamine a titration with *o*-nitroaniline as indicator can be carried out.

Titration with lithium aluminium dibutylamide has been suggested by Small[24] and by Jordan.[25] The rather complicated procedure is described below (recommended Method *D*, p. 116).

Baumann and Simon[26] described the titration of potassium salts of very weak acids (alcohols, sterically hindered phenols) in liquid ammonia at −34°C, using a glass electrode as indicating electrode and a cadmium/cadmium nitrate reference electrode. Due to the very low autoprotolytic constant of ammonia ($\sim 10^{-29}$) the span covered in ammonia is at least 13 p*K*-units larger than in water. This means that acids with p$K_a$-values (in water) between 12 and 39 can be differentiated in ammonia.

The compound is dissolved in liquid ammonia and measured lengths of a thread of potassium, about 1 mm in diameter, are added, the variation in potential being followed as usual. It is also possible to prepare a solution of potassium amide in ammonia, dissolve the sample of hydroxy compound in this solution, then back-titrate the excess potassium amide by addition of ammonium chloride in increments of $17 \pm 1$ mg.

CHAPTER 10

# INDIRECT ESTIMATION OF HYDROXY COMPOUNDS

## 1. Acylation Methods

### (*a*) *Acetylation with Acetic Anhydride*

(i) *With perchloric acid as catalyst.* For the estimation of alcohols and phenols, Fritz and Schenk[27] and Schenk and Fritz[28] recommended acetylation with an excess of a solution that is 2 M with respect to acetic anhydride in ethyl acetate (or pyridine[28]) and 0·16 M with respect to perchloric acid. Alternatively, this acetylating mixture can be diluted with 7 volumes ethyl acetate, thus making it 0·25 M in acetic anhydride and 0·02 M in perchloric acid. Excess acetic anhydride is hydrolyzed by addition of water, and the acetic acid is determined by titration with 0·55 M sodium hydroxide in water/methyl cellosolve or with 0·15 M sodium hydroxide in water. (For details, see recommended Method *E* (*a*), p. 117).

Klouček *et al.*[29] confirmed the results of Fritz and Schenk[27] but found the pyridine method[28] not quite satisfactory. They used the ethyl acetate method for semi-micro estimation and found it sufficiently accurate also for sterically hindered alcohols and phenols.

Magnuson and Cerri[30] preferred a solution of acetic anhydride in 1,2-dichloroethane containing perchloric acid for the estimation, the advantage being that the acetylating reagent can be prepared without cooling (for the ethyl acetate reagent the temp should be kept below 5°C during the preparation). The reagent is colourless or very pale yellow, and it is stable for at least 2 months.

According to Jordan[31] hydroxyl groups in high molecular weight alcohols and complex organic mixtures can be determined in less than 10 min at room temp by the Fritz and Schenk[27] method (perchloric acid concentration 0·05–0·45 M). Acetylation of polyethylene oxide adducts of straight chain monohydric alcohols proceeds normally and without degradation during the first 10 min, but with extended reaction time partial degradation of the polyethylene oxide chain takes place, amounting to nearly 8% after 30 min.

(ii) *With p-toluenesulphonic acid as catalyst.* For the estimation of polyoxyethylene polyols Madeline and Debourg(32) as well as Stetzler and Smullin(33) preferred acetylation with acetic anhydride in ethyl acetate containing *p*-toluenesulphonic acid as catalyst. This catalyst has the advantage over perchloric acid in that it does not attack the ether groups so that degradation of the chain is avoided. Pietrzyk and Belisle(34) suggested 2,4-dinitrobenzenesulphonic acid as catalyst and compared the catalytic effect of this acid and of *p*-toluenesulphonic acid.

The procedure is given below (recommended Method *E* (*b*), p. 118).

(iii) *With pyridine as catalyst and reaction partner.* The oldest procedure for the estimation of hydroxy compounds by acetylation was recommended by Verley and Bölsing.(35) The hydroxy compound is refluxed with excess of a mixture of one part acetic anhydride with 10 parts pyridine, the excess anhydride hydrolyzed and the acetic acid (and pyridinium acetate) titrated with standard sodium hydroxide. Delaby and Sabetay(36) modified the reagent to a mixture of one part acetic anhydride with two parts pyridine and used only 200% excess of acetic anhydride (against originally 400%), but Hauschild *et al.*(37) found this excess insufficient to ensure quantitative acetylation, recommending a procedure closely similar to that suggested by Verley and Bölsing.(35) (For details, see recommended Method *E* (*c*), p. 119.)

According to Quattrone, Jr. and Choy(38) the acetic anhydride–pyridine method allows the quantitative determination of down to 100 μg of primary alcohols in mixtures containing also secondary and tertiary alcohols and water when the back-titration is followed potentiometrically. (For details, see recommended Method *E* (*d*), p. 119.)

Benson and Turner(39) acetylated with acetic anhydride labelled with $^{14}C$ to an activity of approx. 2 μc per m-mole, using one of the above-mentioned pyridine methods for the acetylation of the hydroxy compound. They isolated the activated ester and determined its activity, and claimed that this method offers accuracy and freedom from errors and that it is applicable to milligram quantities of the compound.

The labelled anhydride is prepared in 90% yield from fused labelled sodium acetate and acetyl chloride by refluxing the mixture for 24 h in dry ether. After fractionation, the product is kept under nitrogen.

Kartha and Kidwai(40) recommended a 20% solution of acetic anhydride in toluene as acetylating reagent for semi-micro and micro determination of hydroxy compounds. The sample is refluxed for 4 h with the reagent, excess acetic anhydride hydrolyzed by boiling with water, and the hot solution titrated with 0·5 N potassium hydroxide with phenolphthalein as indicator. A blank is run in parallel.

*(b) Acetylation with Acetyl Chloride*

Smith and Bryant[41] recommended a 1·5 M solution of acetyl chloride in toluene for the estimation of hydroxy compounds. The recommended method is described below (Method *E* (*e*), p. 120).

*(c) Propionylation*

Whereas the acetylation methods usually fail for the determination of tertiary alcohols, Pesez[42] claimed that even tertiary alcohols may be quantitatively propionylated with a reagent prepared by mixing 30 ml glacial acetic acid with 5 ml propionic anhydride and dissolving 1 g *p*-toluenesulphonic acid (not contaminated with sulphuric acid) in the mixture. (For details, see recommended Method *E* (*f*), p. 120.)

*(d) Stearoylation*

Sully[43] found acylation with stearic anhydride a convenient method for the estimation of hydroxy compounds. (For details, see recommended Method *E* (*g*), p. 121.)

*(e) Acylation with 3,5-Dinitrobenzoyl Chloride*

Robinson, Jr. *et al.*[44] acylated primary and secondary alcohols, polyols, sugars and phenols with 3,5-dinitrobenzoyl chloride in pyridine and determined the excess acid chloride after hydrolysis with water. As in the other acylation methods, primary and secondary amines as well as oximes interfere. The authors claimed that by employing a longer reaction time than the standard 15 min, tertiary alcohols may be determined in a much wider range than by other acylation methods. (For details, see recommended Method *E* (*h*), p. 121.)

Instead of titrating hydrochloric acid and 3,5-dinitrobenzoic acid after hydrolysis Robinson, Jr. *et al.*[45] as well as Smith and Haglund[46] recommended the estimation of the isolated 3,5-dinitrobenzoate by titration of its non-aqueous solution with tetrabutylammonium hydroxide:

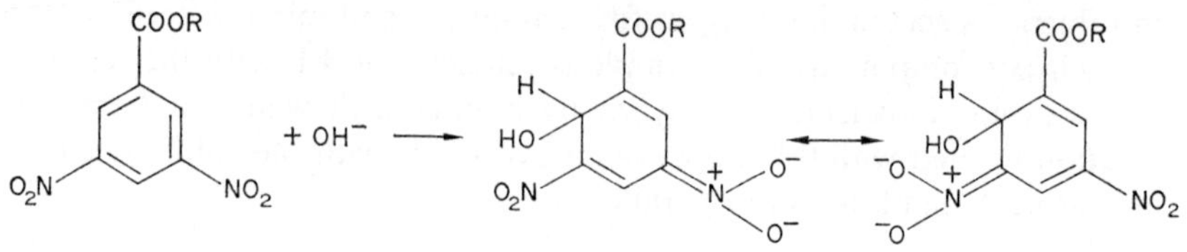

(For details, see recommended Method *E* (*h*), pp. 121–122.)

### (*f*) *Acylation with Anhydrides of Dibasic Acids*

As indicated in Part B, 5.6 (*f*), p. 49, acid esters of alcohols, including many tertiary alcohols, are formed quantitatively by reaction between the hydroxy compound and the anhydride of a dibasic acid. When the acid ester has been isolated its equivalent weight may be determined by titration with 0·1 N sodium hydroxide. For acid phthalates $E_{\text{Alcohol}} = E_{\text{Ester}} - 148{\cdot}1$, for acid tetrachlorophthalates $E_{\text{Alcohol}} = E_{\text{Ester}} - 285{\cdot}9$.

Siggia *et al.*(47) found that the dianhydride of pyromellitic acid reacted so rapidly with alcohols that the esterification rate is comparable to that of acetic anhydride catalyzed by perchloric acid. Kingston *et al.*(48) found that the esterification reaction is strongly catalyzed by imidazole so that quantitative acylation is obtained within 5 min. (For details, see recommended Method *E* (*i*), p. 122.)

### (*g*) *Acylation with Nitric Acid in Acetic Anhydride*

For the estimation of primary alcohols, Schenk and Santiago(49) recommended formation of alkyl nitrates by reaction of the alcohol with a solution of nitric acid and acetic anhydride in acetonitrile. The esterification is complete in 20 min at room temp; acetic anhydride (and acetyl nitrate) is then hydrolyzed with water, and the excess acid titrated with 0·6 N sodium hydroxide in water/methanol. (For details, see recommended Method *E* (*j*), p. 123.)

### (*h*) *Acylation with the Anhydride of Methyl Hydrogen Oxalate*

Oppenauer(48a) claimed that even difficultly esterified alcohols of all three types reacted with the anhydride of methyl hydrogen oxalate, and described methods for the preparation of the reagent and for its use.

$$H_3COOCCOOCH_3 + CH_3COOK \rightarrow H_3COOCCOOK + CH_3COOCH_3$$

Reflux, for 3 h, 1170 g dimethyl oxalate with 1000 g anhydrous potassium acetate and 180 ml water in 2 l methanol. Leave the mixture for 2 days at 5–8°C; isolate the precipitate by suction on coarse filter cloth and wash with 500 ml ice-cold methanol. From the mother liquor a further crop may be obtained after neutralization with potassium carbonate. Total yield: 1205–1286 g = 86–91%.

$$H_3COOCCOOK + SOCl_2 \rightarrow H_3COOCCOCl + KCl + SO_2$$

To a 3-necked flask equipped with reflux condenser and Hirschberg stirrer add 850 ml thionyl chloride. Then add over a period of 30 min 1240 g of methyl potassium oxalate; then slowly raise the temperature to 100°C (oil or water bath) and reflux until the evolution of sulphur dioxide has ceased (about 4 h). Cool and add 500 ml dry ether. Collect the precipitated potassium chloride on coarse filter cloth and wash with 1000 ml dry ether. Fractionate the filtrate and washings using a 50-cm Raschig column. Yield: 1029 g = 97%. B.p. 117–118°C.

$$H_3COOCCOOK + ClOCCOOCH_3 \rightarrow H_3COOCCO—O—OCCOOCH_3 + KCl$$

Place 150 g methyl potassium oxalate in a 3-necked flask (1000 ml) equipped with reflux condenser, separatory funnel, and Hirschberg stirrer. Add 350 ml benzene and distil off 150 ml to remove water completely. Cool to 50°C and add through the separatory funnel a solution of 122 g oxalic ester chloride in 150 ml benzene over a period of several min, then reflux the mixture for 2–3 h. Cool to 50°C, filter on coarse filter cloth, and wash the filter cake with 200 ml benzene heated to 50–60°C. Chill slowly over a period of several h and for up to 2 d, and seed with crystals from a preceding preparation. Another crop may be obtained from the mother liquor after concentration. Total yield: 180 g = 95%. M.p. 65–67°C. Solubility in benzene 1·5% at 5°C; miscible with benzene at higher temp.

$$H_3COOCCO—O—OCCOOCH_3 + ROH \rightarrow H_3COOCCOOR + HOOCCOOCH_3$$

Shake until homogeneous in a chilled 100-ml separatory funnel 0·1–1 m-mole of an alcohol with 0·1–0·2 M solution of the anhydride in benzene, then add a slight excess of pyridine. Leave the mixture to stand for 15 min at 5–10°C. Add 10–20 ml water, shake the mixture for several sec, transfer the aqueous phase to another separatory funnel, and shake it with benzene. The combined benzene phases are washed twice with water and the combined aqueous phases are titrated with 0·1 N sodium hydroxide, indicator phenolphthalein.

Carry out a blank, then calculate the equivalent weight of the ester from the difference between blank and experiment.

$$\text{Then } E_{\text{Alcohol}} = E_{\text{Ester}} - 86{\cdot}0.$$

### (i) *Acylation with p-Toluenesulphonyl Chloride*

Mesnard *et al.*[48b, 48c] indicated that *p*-toluenesulphonyl chloride in pyridine reacted normally with primary and secondary alcohols, glycols, phenols, and hydroxyethylcellulose but did not react with tertiary alcohols.

The reaction is not complete, however, with some secondary alcohols and phenols.

The method is recommended for the determination of tertiary alcohols in essential oils. The total alcohol content is determined by acetylation with acetic anhydride–pyridine *ad modum* Delaby and Sabetay[(36)] (see p. 78); the contents of primary and secondary alcohols are as follows:

Weigh accurately 0·2–0·3 g of the essential oil and add 5 ml of a 20% solution of *p*-toluenesulphonyl chloride in pyridine. Leave the mixture for 1·5 h at room temp; then add 10 ml pyridine and 30 ml water and titrate with 0·5 N sodium hydroxide, indicator phenolphthalein. Compare with a blank; then calculate the pyridine-tosylation index, i.e. weight (mg) of *p*-toluenesulphonic acid combined with 1 g of the hydroxyl-containing compound, and compare it with the pyridine-acetylation index. From the difference, the amount of tertiary alcohols may be calculated.

### (*j*) *Acylation with o-Sulphobenzoic Anhydride*

Iyer and Mathur[(48d)] and Narang *et al.*[(48e)] recommended esterification with *o*-sulphobenzoic anhydride for the microdetermination of hydroxyl groups in primary and secondary alcohols ($C_1$–$C_{18}$), glycols, hydroxy acids, and polyglycol ethers.

*Reagent.* 0·5 M solution of *o*-sulphobenzoic anhydride in dioxan.

*Procedure.* In a pyrex glass weighing-tube with B-14 ground glass stopper weigh accurately 0·1–0·2 m-equiv. of the hydroxy compound, add excess (100–150 mole%) of the reagent solution and reweigh the tube. Add 2 ml pyridine, stopper the tube, and fit a rubber band to hold the stopper in position when pressure develops upon heating. Heat in a water-bath for 15 min, cool, then transfer and wash the contents of the tube into a 100-ml Erlenmeyer flask. Titrate with 0·1 N sodium hydroxide, indicator phenolphthalein. Compare with a blank, and calculate the equivalent weight of the hydroxy compound from the difference between blank and experiment.

The activity of *o*-sulphobenzoic anhydride is significantly enhanced by the addition of pyridine, presumably due to the formation of an *N*-acyl-pyridinium compound. Compounds of this type are known to be strong acylating reagents.

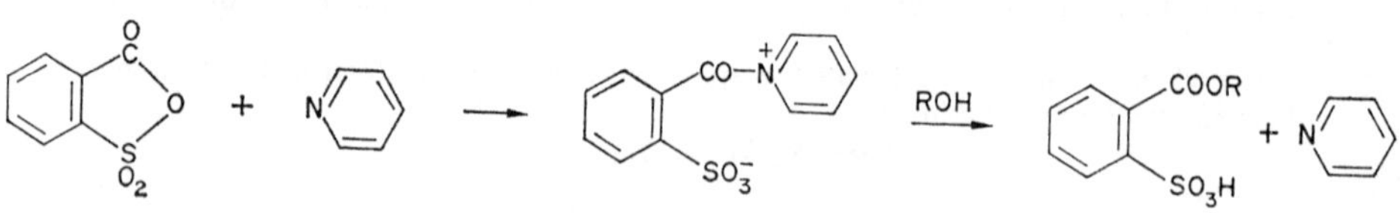


## 2. Etherification Methods

When 2,4-dinitrophenyl ethers have been prepared for the characterization of hydroxy compounds as indicated in Part B, 6.1 (*b*), p. 55, the ethers may be used[46] for the determination of the equivalent weight of the hydroxy compound.

Dissolve 0·1–0·2 g of the ether in 25 ml acetone and titrate the solution potentiometrically (glass electrode/calomel electrode) with 0·1 N tetrabutylammonium hydroxide in benzene/methanol.

From the equation:

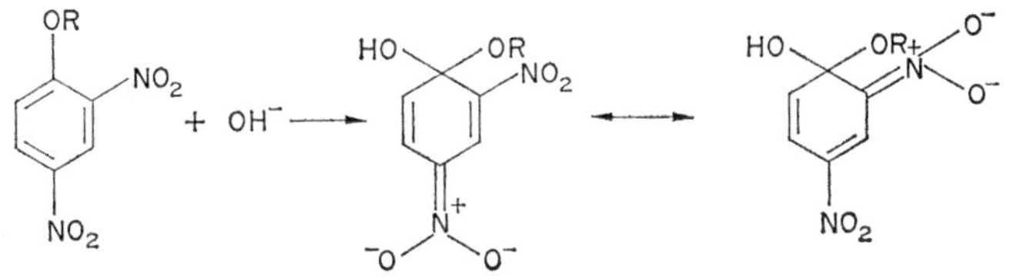

it is seen that $E_{Phenol} = E_{Ether} - 182{\cdot}1$.

### 3. Estimation through Phenylurethane Formation

Several authors (e.g. Dreher,[50] Reed *et al.*,[51] Otey *et al.*[52]) have recommended reaction with a standardized solution of an arylisocyanate to form an arylurethane for the estimation of polyethers (polyoxyethylene compounds). Excess isocyanate is determined either through back-titration with a standardized solution of diisobutylamine, or by addition of excess diisobutylamine and back-titration of the excess with standardized hydrochloric acid, using bromophenol blue as indicator in both instances. Lower alcohols, but not phenols, can also be estimated in this way. (For details, see recommended Method *F*, p. 123.)

### 4. Bromometric Estimation of Hydroxy Compounds

(*a*) *Phenols*

Several procedures have been recommended for the estimation of phenols bromometrically. These are based on the rapid substitution of bromine for hydrogen atoms in the *ortho-* or *para-*position to the hydroxyl group:

It is seen that 6, 4, or 2 equivalents of bromine are used up, depending on the number of *o*- and *p*-hydrogen atoms.

Two sources of error should be considered. (i) If too great an excess of bromine is used formation of "tribromophenol bromide" may take place, and in the back-titration of excess bromine difficulty in determination of the end-point will arise, as the tribromophenol bromide will react slowly with hydriodic acid, liberating iodine. (ii) Some substituents, e.g. carboxyl groups, are so easily exchanged with bromine that compounds such as salicylic acid and *p*-hydroxybenzoic acid will use up six equivalents of bromine in the reaction.

Procedures for carrying out the estimation have been suggested by Veibel,[(53)] Smith,[(54)] Chikryzova and Khomenko,[(55)] Schoenberger,[(56)] Karpov[(57)] and Huber and Gilbert.[(58)] Details are given in recommended Methods *G* (*a*), *G* (*b*), *G* (*c*), *G* (*d*), *G* (*e*) and *G* (*f*), pp. 124–126.

### (*b*) *Aliphatic Hydroxy Compounds*

Konishi *et al.*[(59)] have developed an iodometric micromethod for the estimation of primary and secondary alcohols by oxidation with an aqueous solution of bromine chloride, making possible the determination of 3–10 μ-moles of primary alcohols, or 8–15 μ-moles of secondary alcohols. From the schemes:

$$RCH_2OH + 2BrCl + H_2O \rightarrow RCOOH + 2HBr + 2HCl$$

$$R_2CHOH + BrCl + H_2O \rightarrow R_2CO + HBr + HCl$$

it is seen that primary alcohols use up four equivalents of the oxidizing agent, and secondary alcohols require two equivalents. Details are given in recommended Method *G* (*g*), p. 126.

Bradley and Penketh[(60)] have developed an iodometric method for the estimation of tertiary alcohols. The alcohol is converted into alkyl iodide by reaction with hydriodic acid, the alkyl iodide is extracted into cyclohexane, then hydrolyzed (or hydrogen iodide eliminated) with potassium hydroxide and finally the hydrogen iodide is converted into iodine by reaction with iodic acid. The authors recommended a spectrometric determination of iodine by measuring the absorbance at 525 nm, but titration with sodium thiosulphate should also be possible. For details, see recommended Method *Q* (*b*), p. 140.

Cheng[(61)] indicated a micromethod for determining adjacent hydroxyl groups in organic compounds, involving the digestion of the sample with hydriodic acid in a sealed tube, each set of adjacent hydroxyl groups producing 1 mole of iodine:

$$R{-}CHOH{-}CHOH{-}R' + 2HI \rightarrow RCH_2OH + R'CH_2OH + I_2$$

Evidently, this reaction may be followed by further reaction of the primary alcohols with hydriodic acid:

$$RCH_2OH + R'CH_2OH + 4HI \rightarrow RCH_3 + R'CH_3 + 2H_2O + 2I_2,$$

thus invalidating the estimation if the second reactions are not quantitative. The author mentioned that two adjacent hydroxyl groups will produce 1 mole of iodine, i.e. he disregarded the second reaction. If, on the other hand, the second reaction is quantitative, then three molecules of iodine will be produced per molecule containing two adjacent hydroxyl groups.

## 5. Oxidation Methods

Oxidation with chromic acid or with vanadic acid has been recommended for the estimation of alcohols but it is difficult to indicate a general procedure because different alcohols behave differently. Procedures for chromic acid oxidation have been given by Jaulmes and Mestres,[(62)] and for vanadium(V) oxidation by Panwar and Gaur.[(63)]

A more specific method for the oxidation of polyols involves oxidation with cerium(IV) sulphate dissolved in aqueous perchloric acid, catalyzed by a silver–manganese catalyst (Guilbault and McCurdy, Jr.[(64)]). Details of this procedure are given below in recommended Method *H* (*a*), p. 127.

Pesez and Bartos[(65)] have developed a colorimetric method for the estimation of primary alcohols by oxidation with ruthenium tetroxide, which they claim to be superior to osmium tetroxide. A disadvantage is that the recommended reagent can be kept for only 2 h. Amounts of alcohol down to 10–30 μg can be estimated. Interference from secondary alcohols is negligible.

Srivastava and Saxena[(66)] described the estimation of glycols in amounts down to 0·1 mg with gold(III) chloride in alkaline solution. The gold(III) ions were reduced to gold and the excess reagent was determined by addition of a known excess of standard potassium hexacyanoferrate(II), the excess of which is determined by titration with standard cerium(IV) sulphate solution, indicator *N*-phenylanthranilic acid. Details are given in recommended Method *H* (*b*), p. 128.

A very promising method for the estimation of primary and secondary alcohols in aqueous solution has been described by Jaselskis and Warriner.[67] The oxidation is carried out with a solution of xenon trioxide prepared from xenon tetrafluoride by reaction with water, which according to Williamson and Koch[68] causes a disproportionation of xenon tetrafluoride to xenon trioxide, xenon, and oxygen. The oxidation of alcohols leads to the formation of carbon dioxide, water, and xenon:

$$CH_3CH_2OH + 2XeO_3 \rightarrow 2CO_2 + 3H_2O + 2Xe$$

$$CH_3CHOHCH_3 + 3XeO_3 \rightarrow 3CO_2 + 4H_2O + 3Xe$$

Excess xenon trioxide is determined iodometrically after adding sodium iodide and acidifying the solution. The range of hydroxyl concentration over which the method can be applied includes both the micro and the semi-micro region; amounts as low as 25 μg can be determined with a relative standard deviation of 4%, and at the 100 μg level the standard deviation is below 1%. (For details, see recommended Method *H*(*c*), p. 128.)

An oxidation of dihydric phenols with the free radical 1,1-diphenyl-2-picrylhydrazyl has been suggested by Blois[69] and has been followed up by Papariello and Janish[70] (who extended the scope of the method to include monohydric phenols) and by Schenk and Brown.[71] The principle of the method is that the free radical shows a strong absorbance of 517 nm, which disappears when hydrogen is taken up by the radical. Whereas Blois[69] indicated that the absorbance is independent of pH from pH 5 to pH 6·5 and therefore recommended an acetate buffered solution, Schenk and Brown[71] found that water affected the rate of reaction with all three dihydric phenols and the stoichiometry of the pyrocatechol oxidation. They therefore recommended the use of absolute ethanol as solvent, but resorcinol then reacts at a negligible rate. For pyrocatechol full reaction is obtained after 50 min, for hydroquinone after 140 min. The reaction is:

$$2(C_6H_5)_2N{-}NC_6H_2(NO_2)_3 + HOC_6H_4OH \rightarrow 2(C_6H_5)_2N{-}NHC_6H_2(NO_2)_3 + OC_6H_4O$$

(For details, see recommended Method *H* (*d*), p. 129.)

## 6. Estimation through Acylation or Dehydration and Karl Fischer Titration of Eliminated Water

Bryant *et al.*[72] found that boron trifluoride catalyzed a reaction between acids and alcohols, leading either to esterification (primary and many secondary alcohols) or to the formation of an alkene (tertiary and some secondary alcohols). In both reactions a molecule of water is eliminated per molecule of alcohol:

$$ROH + CH_3COOH \xrightarrow{BF_3} CH_3COOR + H_2O$$

$$\begin{matrix} R^1 \\ R^2 \end{matrix}\!\!>\!C(OH)CH_2R^3 \xrightarrow{H^+,\ BF_3} \begin{matrix} R^1 \\ R^2 \end{matrix}\!\!>\!C{=}CHR^3 + H_2O$$

The alcohol can thus be determined by carrying out a Karl Fischer titration of water (see recommended Method *I*, p. 129).

Petrova and Novikova(73) used a similar method to estimate tertiary alcohols (and primary and secondary alcohols that dehydrate readily) in the presence of stable alcohols by dissolving an exactly weighed sample in toluene, and adding *p*-toluenesulphonic acid as catalyst. After refluxing an aliquot for 5 min, the sample after cooling is titrated with Karl Fischer reagent. A blank is run on the solvent to which has been added the same amount of catalyst as in the experiment. For pure alcohols the equivalent weight is calculated as usual, one mole of water being eliminated per dehydrated alcohol group. For mixtures of stable and dehydratable alcohols, only the percentage of the latter can be calculated.

## 7. Estimation of Active Hydrogen in Hydroxy Compounds with Grignard Reagent or Lithium Aluminium Hydride

For the estimation of compounds containing active hydrogen, e.g. alcohols, Zerevitinov(74) recommended a reaction with methyl magnesium iodide, collecting the methane evolved in an azotometer and correcting the volume for ether vapours. This method was developed by several authors, e.g. Zaugg and Lauer,(75) who used the apparatus shown in Fig. 3. With this apparatus a correction for ether vapours is also necessary. Terent'ev and Kirejeva(76) introduced a simplification by collecting the gas liberated over 50% ethanol, which retains the ether. The remaining methane is then collected over water in an azotometer, and the volume has only to be corrected for the vapour pressure of water at the temperature of the azotometer.

Instead of methyl magnesium iodide, a solution of lithium aluminium hydride may be used for the estimation of active hydrogen. With this reagent hydrogen and not methane is evolved.

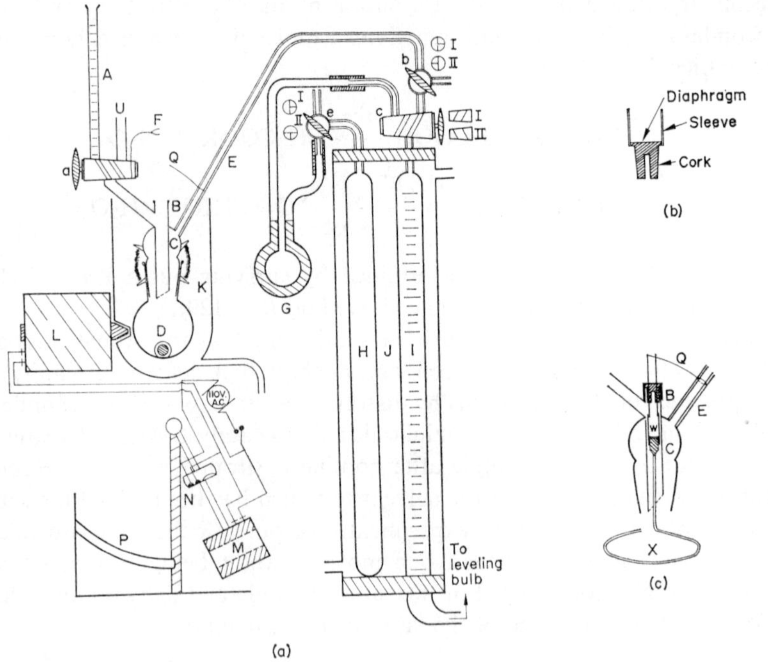


FIG. 3. Determination of active hydrogen according to Zaugg and Lauer.[75] (a) Diagram of apparatus; (b) Rubber sleeve serum stopple; (c) Serum stopple placed at B during a determination. (Reprinted from *Anal. Chem.* **20**, 1022 (1948). Copyright (1948) by the American Chemical Society. Reprinted by permission of the copyright owner and the authors.)

Both reagents are used as 1–2 *N* solutions in ether (Grignard reagent), diisopropyl ether, or tetrahydrofuran (lithium aluminium hydride).

Lithium aluminium hydride has proved to be extremely useful for the determination of hydroxyl groups in compounds containing α-epoxide groups, as these are attacked and opened by the reagents usually applied for the estimation. Stenmark and Weiss[77] claimed that such α-epoxide groups are not attacked by lithium aluminium hydride, which has, besides, the advantage that the reaction is more vigorous than that with the Grignard reagent, so that even severely hindered phenols will react quantitatively. They recommended an apparatus shown in Figs. 4 and 5.

The method has been studied also by Falgoux *et al.*,[78] who determined

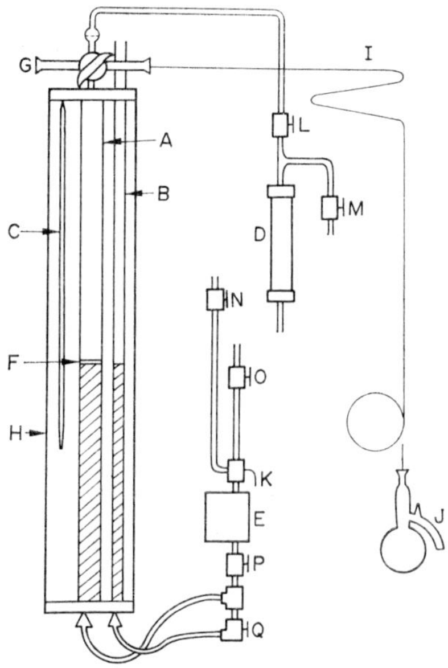


FIG. 4. Determination of hydroxyl groups with lithium aluminium hydride (Stenmark and Weiss[77]).

Schematic diagram of apparatus

A. 100-ml burette
B. Levelling tube
C. Thermometer
D. Nitrogen dryer containing Dehydrite, to nitrogen source
E. Mercury reservoir, 250 ml
F. Tetrahydrofuran layer
G. Vent port
H. Water-filled burette jacket
I. Stainless steel Hypotubing, 14-gauge
J. Reaction flask
K. Three-way valve control for mercury lift steel needle valve
L. Nitrogen flushing
M. Control to tetrahydrofuran dispenser
N. Air supply
O. Vacuum supply
P. Burette
Q. Levelling tube

(Reprinted from *Anal. Chem.* **28**, 1784 (1956). Copyright (1956) by the American Chemical Society. Reprinted by permission of the copyright owner and the authors.)

the hydroxy group content of epoxy resins by measuring the volume of hydrogen evolved when an accurately weighed sample reacted with a solution of lithium aluminium hydride in tetrahydrofuran contained in a Roth gas-burette. Water in the solvent was determined by titration with Karl Fischer reagent, acids by titration with standard sodium hydroxide,

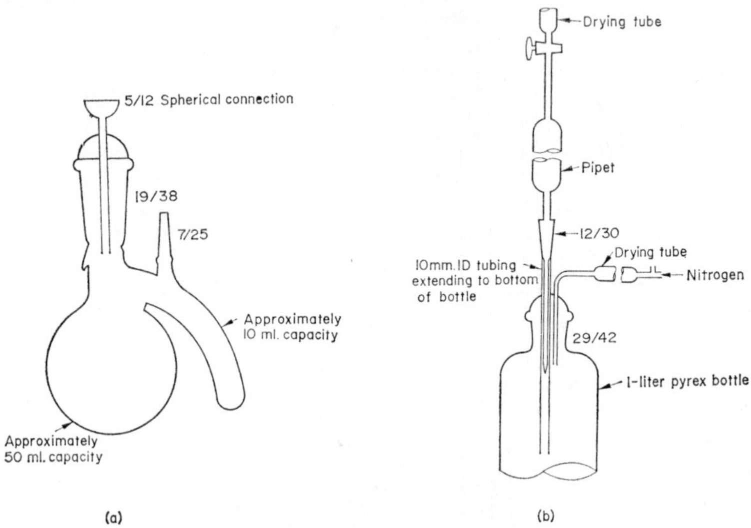


FIG. 5. Details of the apparatus in Fig. 4. (a) Reaction flask assembly; (b) Storage and dispensing system for tetrahydrofuran. (Reprinted from *Anal. Chem.* **28**, 1784 (1956). Copyright (1956) by the American Chemical Society. Reprinted by permission of the copyright owner and the authors.)

and hydroxyl-containing stabilizers by u.v. spectroscopy, the corresponding corrections being introduced in the results obtained.

Norton *et al.*(79) used an improved version of a gas chromatographic method for the determination of active hydrogen with lithium aluminium hydride as suggested by Chumachenko *et al.*(80), applying nitrogen as carrier-gas instead of argon, molecular sieves as stationary phase instead of activated charcoal, and molecular sieves for drying both the carrier-gas and the tetrahydrofuran used as solvent.

Details of the method recommended by Norton *et al.* are given in recommended Method *J*, p. 129.

## 8. Determination of the Equivalent Weight of Hydroxy Compounds through Derivatives Mentioned in Part B

*(a) Determination of Phenols through Aryloxyacetic Acids*

The aryloxyacetic acids prepared as indicated by Koelsch,(81) see Part B, 6.1 (*d*), p. 38, are sufficiently strong acids to allow their titration with standard sodium hydroxide and phenolphthalein indicator. The equivalent weight, *E*, of the phenol is found by subtraction of 58·0 from that of the aryloxyacetic acid.

(*b*) *Determination of the Equivalent Weight of Tertiary Alcohols through Alkylthiouronium Picrates or 3,5-Dinitrobenzoates*

These derivatives were recommended in Part B, 6.2, p. 39, for the characterization of tertiary alcohols. Schotte and Veibel(82) dissolved an accurately weighed sample of the picrate in 20 ml glacial acetic acid and titrated the solution with 0·1 N perchloric acid in glacial acetic acid either potentiometrically (glass electrode/calomel electrode) or visually with 1–2 drops of 1% crystal violet in glacial acetic acid as indicator.

The equivalent weight of the picrate is calculated as usual and then $E_{\text{Alcohol}} = E_{\text{Picrate}} - 287{\cdot}2$.

Jureček *et al.*(83) preferred the 3,5-dinitrobenzoates as they are more easily purified from 3,5-dinitrobenzoic acid than are the picrates from picric acid. The 3,5-dinitrobenzoates can be titrated with perchloric acid just as for picrates. Then $E_{\text{Alcohol}} = E_{\text{Salt}} - 270{\cdot}2$.

A titanometric estimation of these derivatives is also possible.

(*c*) *Determination of the Equivalent Weight of Alcohols through the Potassium Alkyl Xanthates*

Potassium alkyl xanthates, prepared as described in Part A, 1.1 (*e*), p. 5, can be used for the determination of the equivalent weight of alcohols. Berger(84) dissolved an accurately weighed sample (0·15–0·20 g) in 20 ml glacial acetic acid and titrated this solution with 0·1 N perchloric acid in glacial acetic acid, potentiometrically (glass electrode/calomel electrode) or visually with 1–2 drops of 1% crystal violet in glacial acetic acid as indicator. The equivalent weight of the xanthate is calculated as usual and then $E_{\text{Alcohol}} = E_{\text{Xanthate}} - 114{\cdot}2$.

Whitmore and Lieber(85) determined the xanthates iodometrically, dissolving an accurately weighed sample (0·15–0·20 g) of the xanthate in 100 ml water, adding a few ml starch solution and titrating with 0·1 N iodine in potassium iodide until a clear blue colour, persisting for some minutes, was obtained.

The reaction scheme:

$$2\text{ROCSSK} + \text{I}_2 \rightarrow (\text{ROCSS})_2 + 2\text{KI}$$

shows that the equivalent weight of the xanthate can be calculated as usual and then $E_{\text{Alcohol}} = E_{\text{Xanthate}} - 114{\cdot}2$.

(d) *Titanometric Determination of the Equivalent Weight through Nitro Group-containing Derivatives*

In Part B were mentioned *m*-nitrophenylurethanes (4.3, p. 25), *p*-nitrobenzoates (5.3, p. 29), 3,5-dinitrobenzoates (5.4, p. 30), and 2,4-dinitrophenyl ethers (6.1 (*b*), p. 37). The equivalent weight, *E*, of all these derivatives can be determined titanometrically and the molecular weight is then 6 *E* for mono-nitro compounds and 12 *E* for dinitro compounds. The molecular weight of the alcohol is then found by subtracting 164·1, 149·1, 194·1 or 166·1, respectively, from the molecular weight of the derivative.

It is seen that an error of, say, 0·5% in the determination of *E* will imply a considerably greater error in the calculated molecular weight of the alcohol; the weight of the alcohol, ROH, is usually only a small part of the weight of the derivative, and the error is concentrated on this small part.

A convenient technique for titanometric estimation of nitro compounds has been described by Kolthoff and Robinson.[86] Details are given in recommended Method *K*, p. 133.

## 9. Polarographic Estimation of Hydroxy Compounds

Vodzinskii and Semchikova[87] studied the anodic polarographic waves on graphite electrodes for a long series of phenols, $RC_6H_4OH$, and indicated the half-wave potentials. For *meta*- and *para*-substituted phenols the half-wave potentials obey a Hammett plot. Values for the half-wave potentials are cited in *Chemical Abstracts*.

Fürst[88] made polarographic estimations of 3,5-dinitrobenzoates or 2,4-dinitrophenyl ethers of hydroxy compounds in alcoholic solution. He claimed that in suitable basic solution the estimation could be carried out without separation of excess reagent.

Silkina *et al.*[89] determined acetylenic tertiary alcohols

$$HC{\equiv}C{-}C(OH)RR'$$

polarographically in 2–4% aqueous potassium hydroxide and found that diffusion currents and differential peak currents were linear functions of the concentration of the alcohol over the concentration range 0–0·1 M.

## 10. Complexometric Methods

(*a*) *Complexes with Bis(8-hydroxyquinolinyl) Vanadate*

Tanaka,[90] Pesez and Bartos,[91] and Stiller[92] made use of a reaction between the vanadium–oxine complex and alcohols for a spectrophotometric micro-analytical estimation of alcohols, the reaction being:[92]

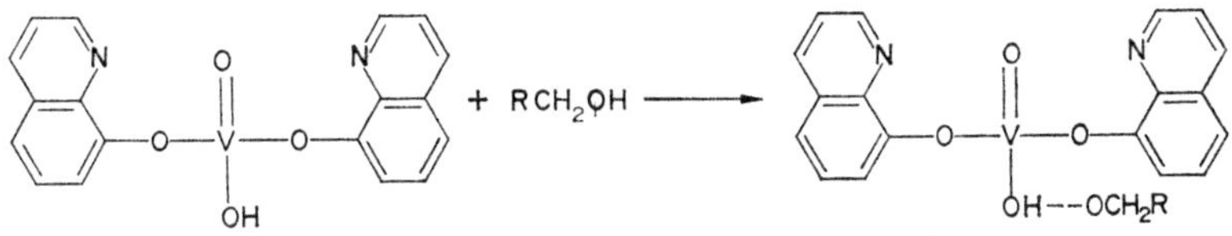


the greyish-green complex (I) shifting colour to red (II), which is more stable than (I).

Complexes (II) are formed with primary, secondary and tertiary alcohols but not with alcohols containing a phenolic or carboxylic group except when the carboxylic group has been modified, e.g. by esterification. Thus, the reaction is negative for lactic acid but positive for lactic acid esters.

The optical density is measured at 390 nm, and compared with a calibration curve for each alcohol.

Details of Stiller's procedure are given below (recommended Method *L* (*b*), p. 136).

Tanaka[90] measured the absorbance at 635 nm, but Pesez and Bartos[91] used 600 nm. For details, see recommended Method *L* (*a*), p. 135.

Later Bartos[93] suggested a fluorimetric estimation of the vanadium oxine complex formed. The red complex was isolated and hydrolyzed in acid medium, the liberated 8-hydroxyquinoline determined by fluorimetry of its magnesium complex by measuring the excitation at 366 nm, and the emission at 520 nm.

(*b*) *Complexes with Titanium(IV) Chloride*

For the spectrophotometric determination of phenolic hydroxyl groups Jacob[94] recommended the formation of their complexes with titanium(IV) chloride. All reagents should be absolutely dry.

Add 1 ml 0·2 M titanium(IV) chloride in dichloromethane (containing 600 μl glacial acetic acid/l to suppress hydrolysis) to 5 ml of a solution of 1–10 mg of the sample in dichloromethane acidified with 25 μl glacial acetic acid; the absorbance of the titanium–chlorophenoxy complex is measured after 3 min at room temp at the proper wavelength (calibration curves necessary) against a blank solution.

Titanium(IV) chloride should be distilled over charcoal, and dichloromethane boiled under reflux for 2 h over potassium carbonate.

### (c) *Determination of Hydroxy Compounds through EDTA-titration*

For the rapid titrimetric determination of polyethylene glycols, Maros *et al.*[95] recommended addition of a known excess of barium tetraphenyl borate to a solution of the polyethylene glycol, removal of the precipitate by filtration, and determination of excess barium ions by titration with EDTA, indicator methylene blue.

Similarly Omboly and Derzsi[96] determined 8-hydroxyquinoline and alkyl- or halogeno-substituted 8-hydroxyquinolines by precipitation at pH 10 with an excess of zinc sulphate, excess zinc being determined complexometrically by titration with EDTA, and Hubicka[97] determined resorcinol in 0·01 M solution (made alkaline with excess ammonia) by precipitation with 0·01 M mercury(II) chloride. The precipitate was isolated by filtration, dissolved in 5 M hydrochloric acid, and mercury(II) determined complexometrically by titration with EDTA.

### (d) *Colorimetric Estimation of Alcohols by Acetylation and Formation of the Corresponding Iron(III) Hydroxamates*

Mizukami and Nagata[98] as well as Gutnikov and Schenk[99] recommended acetylation followed by formation of hydroxamic acid, which was estimated colorimetrically as its complex iron(III) salt. It is claimed that one single calibration curve is sufficient for almost any kind of hydroxy compound and that amino- and carboxyl groups do not interfere. Tertiary alcohols cannot be estimated by this method. (For details, see recommended Method *M*, p. 137.)

### (e) *Complexometric Estimation of Hydroxy Compounds through Complex Formation with Ammonium Hexanitratocerate*

Obtemperanskaya *et al.*[100, 101] described a colorimetric method for estimating water-soluble alcohols, and water-insoluble alcohols and phenols; these compounds formed red–orange complexes with ammonium hexanitratocerate(IV) in nitric acid solution. A drawback is that the complexes are stable for only 3–10 min at pH 0·4–0·9. The sensitivity of the method is about $10^{-5}$ moles of alcohol per ml. Tertiary alcohols can also be estimated. (For details, see recommended Method *N*, p. 138.)

## 11. Fluorimetric Estimation of 1,2-Diols

Pesez and Bartos(102) have established a fluorimetric method for the estimation of 1,2-diols, based on the oxidation of the diol with periodic acid. Aldehydes are formed, and these are then treated with ethyl acetoacetate and ammonium acetate to convert them into a dihydrolutidine, which is estimated by determining the excitation at 366 nm and the emission at 470 nm in a spectrofluorimeter. (Details are given in recommended Method *O*, p. 139.)

## 12. Spectrophotometric Methods

*(a) Free Phenols*

McKinney and Reynolds(103) have developed a method for the photometric titration of phenols, using an u.v.-light source, a monochromator, and a photoelectric microphotometer.

For each phenol to be estimated the ultraviolet and visible absorption spectra are registered and a wavelength at which the phenate anion absorbs is chosen for the titration. The instrument is set initially at zero absorption, at the wavelength selected, by adjustment of the slit width.

Titration of a 0·1 m-equiv. sample of a phenol with 0·05 N alcoholic sodium hydroxide was carried out using 100 ml butylamine solvent in a cubic glass cell of 125 ml capacity. The titrant was added in 0·2 ml increments, and the absorbance values were plotted against titrant volume. Straight lines were drawn through the linear portion of the plots, and the titrant volume measured between the two intersections gave the volume of titrant required for each sample of phenol. During the titration the solution in the cell was stirred continuously, and nitrogen was passed through the compartment.

Phenols with $pK_a$ values in water ranging from 5 to 11 can be titrated without difficulty.

*(b) Derivatives of Hydroxy Compounds*

(i) *p-Nitrobenzoates.* Scoggins(104) determined trace quantities of alcohols by converting them into *p*-nitrobenzoates through reaction with *p*-nitrobenzoyl chloride in pyridine (see Part B, 5.3, p. 30). Cyclohexane was added, the solution washed with 2 N potassium hydroxide, then twice with 2 N hydrochloric acid, repeating these two washings. Finally the solution was diluted to the mark in a volumetric flask and the absorbance at 253 nm was read and compared with the absorbance of a blank. The amount of phenol present was found from a calibration curve.

(ii) *3,5-Dinitrobenzoates.* The conversion of primary and secondary alcohols to 3,5-dinitrobenzoates with 3,5-dinitrobenzoyl chloride and pyridine can be used in different ways for the estimation of μg-quantities of alcohols.

Pesez and Bartos[(105)] dissolved several μg of the sample in 0·1 ml of a mixture of pyridine and benzene (1/1), left this mixture for 20 min at 18–20°C, then added 0·1 ml 0·1 M 3,5-dinitrobenzoyl chloride in pyridine; after 5 min 5 ml of 25% piperazine hexahydrate in formamide was added and the optical density was measured at 510 nm. The red colour develops only for primary and secondary alcohols; phenols give a transient colour.

Obtemperanskaya and Borodina[(106)] used the Janovsky reaction[(107)] in which *m*-dinitro compounds in acetone solution give a colour upon addition of sodium or potassium hydroxide. They dissolved the sample in pyridine and added to an aliquot 1 ml of 10% solution of 3,5-dinitrobenzoyl chloride in pyridine. The mixture was left to stand for 20 min, then 25 ml 2 N hydrochloric acid was added. Shake for 20 min, add 5 ml hexane, and shake again for 20 min. Complete the volume to 50 ml with water. To 2 ml of the hexane layer add 7 ml acetone, then 0·3 ml 2 N sodium hydroxide, and complete the volume to 10 ml with acetone. Measure the absorbance in a 1-cm cell at 410 nm and at 556–560 nm and compare with a calibration curve prepared with known concentrations of a 3,5-dinitrobenzoate.

The sensitivity of the method was claimed to be $10^{-6}$ M of phenol.

(iii) *Alkyl nitrites.* Shmulyakovskii[(108)] and Meluzova *et al.*[(109)] converted primary and secondary alcohols ($C_9$–$C_{18}$) into alkyl nitrites and measured the absorbance at 340–390 nm[(108)] or at 960–1200 nm.[(109)] In u.v. the absorption bands of primary and secondary alcohols show a difference of only 1–2 nm; the values of the extinction, $\epsilon$, however, are different for the two series but within each series are nearly independent of M. Cyclic alcohols behaved similarly to linear secondary alcohols with regard to extinction.

Crummett[(110)] studied the estimation of the primary hydroxyl group content of polypropylene glycol by the Shmulyakovskii method[(108)] but found it unsatisfactory because of the instability of the nitrosation products. By carrying out the reaction in absence of air, by inclusion of a filtration step, and by correcting for background absorbance he succeeded in obtaining satisfactory results.

(iv) *Colorimetric estimation of hydroxy compounds through formation of azo compounds.* Katz and Keeney[(111)] treated alcohols with *p*-phenylazobenzoyl chloride (see Part B, 5.5, p. 30) to form *p*-phenylazobenzoates which were then isolated chromatographically and the molar extinction at 322 nm was determined in hexane solution. The authors indicated that the *p*-phenylazobenzoates of primary alcohols had a molar extinction of 28,600.

Popa *et al.*[112] prepared very pure azo compounds of phenol, e.g. 4-hydroxy-4′-nitroazobenzene, (I) and 4-hydroxyazobenzene-4′-sulphonic acid (II) in order to have standards for the spectrophotometric estimation of phenol. (I) showed maximum absorbance in the violet region, (II) in the blue region. Diazotized *p*-nitroaniline and sulphanilic acid are colourless and excess reagent therefore does not invalidate the estimation.

A spectrophotometric microestimation of monohydric aliphatic alcohols has been described by Hoare *et al.*[113] They converted the alcohol into alkyl nitrite, which then reacted with α-naphthylamine, converting it into α-naphthalenediazonium ion which in turn couples with excess α-naphthylamine, forming an azo dye the absorbance of which was measured at 550 nm in 1-cm cells. The reactions were:

$$ROH + H_2\overset{+}{N}O_2 \rightarrow ROH\overset{+}{N}O + H_2O$$

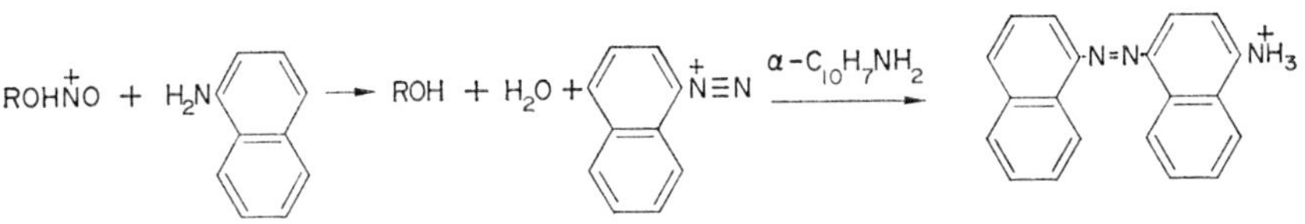

It is seen that all alcohols give the same dye, but the absorbance given by a fixed weight of an alcohol will be inversely proportional to its molecular weight. (Details are given in recommended Method *P*, p. 139.)

(v) *Estimation through condensation with 4-aminoantipyrine.* Etlinger *et al.*[114] and Shaw[115] made use of the Emerson[116, 117] colour reaction of phenols with aminoantipyrine, extracted the coloured substance with chloroform, measured the optical density at 460 nm, and compared the result with a calibration curve.

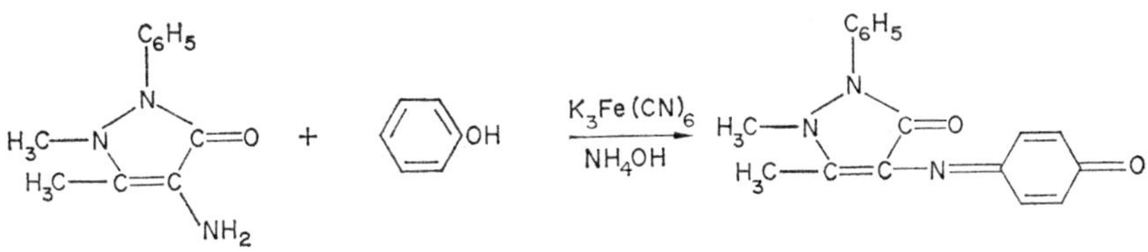

Dissolve 10–20 mg of the compound in 10 ml water to which 2–3 drops of 6 N ammonium hydroxide have been added (if insoluble in water, dissolve the substance in a few drops of ethanol and use this solution). Divide the solution into two equal portions. To one portion, add 2 drops of 2% 4-aminoantipyrine solution, the other portion being kept as a control. To each tube add several drops of 8% potassium hexacyanoferrate(III). A large excess of oxidizing agent should be avoided because it may decolorize the dye formed. After the development of full colour, add 2 ml chloroform to each tube, shake, then separate the chloroform layer into a measuring flask which is filled to the mark with chloroform. The optical density of the two chloroform layers is measured at 460 nm.

(vi) *Estimation of hydroxy compounds by near infrared spectrometry.* Several authors[(118–120)] have described a method for estimating aliphatic and aromatic hydroxy compounds by using the hydroxyl stretching band in the region 2·7–3·0 μ (phenols) or 2·0–3·2 μ (polyethers and polyesters). The instrument used was a Beckman model DK-2 spectrophotometer with special near infrared transparent silica cells. The method is fast and reproducible. It allows the analysis of samples containing as little as 25 ppm of phenolic hydroxyl, and because intramolecular hydrogen bonding causes shifts in the hydroxyl band it is often possible to analyse samples containing several phenolic species.

(vii) *Miscellaneous methods.*

1. Anderson and Zaidi[(121)] determined 1,2-diols by a modified Zeisel method, allowing the simultaneous determination of ethylene (or propylene) together with alkyl iodide by vapour-phase infrared spectroscopy based on the absorbance of ethylene at 955 $cm^{-1}$ and of ethyl iodide at 1215 $cm^{-1}$m, using an apparatus shown in Fig. 6.

Under standard conditions for Zeisel determinations (5 mg sample refluxed with 6 ml hydriodic acid (55%) using nitrogen as flow gas at 6–8 ml per min) ethanediol gives 85 moles per cent of ethylene and 15 moles per cent of ethyl iodide. The evolution of ethylene is extremely rapid. From start of reflux the evolution of ethylene is complete within 10 min; the evolution of ethyl iodide is slower, being complete in 30 min.

2. Scoggins and Miller[(122)] converted tertiary alcohols into alkyl iodides with 50–55% hydriodic acid. Details are given in recommended Method *Q* (*a*), p. 140.

3. Nisli and Townshend[(123)] estimated 1,2 diols by oxidation with periodic acid and spectrophotometrical estimation of the iodine formed by reaction of the iodic acid with hydriodic acid, measuring the absorbance of the triiodide ion at 350 nm. Excess of periodic acid was masked with molybdate.

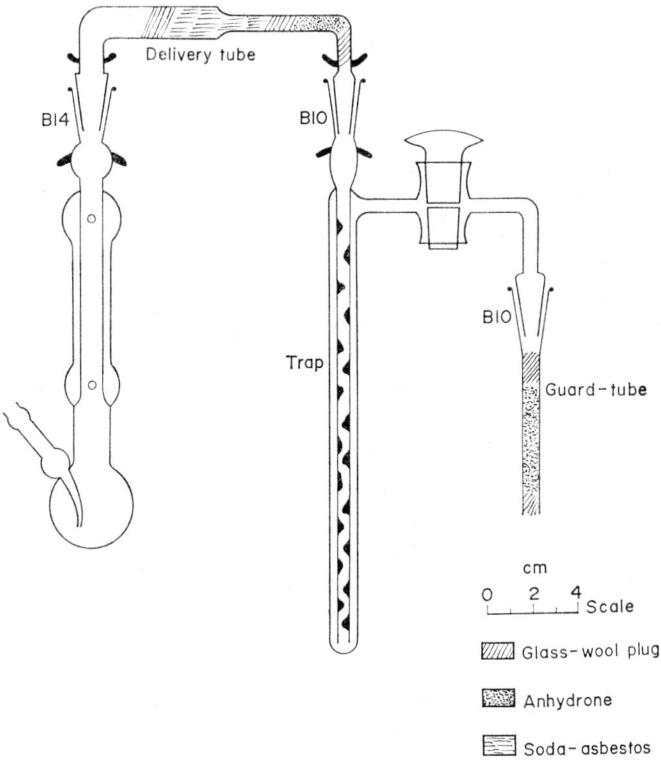


FIG. 6. Apparatus for modified Zeisel determination of 1,2-glycols (Anderson and Zaidi[121]). (Reprinted by courtesy of Pergamon Press Ltd. and the authors.)

## 13. Chromatographic Methods

(*a*) *Paper Chromatography*

Most of the paper chromatographic methods recommended are either only semiquantitative or too complicated to be recommended for general use, although they may be useful as routine methods within special fields. Such methods have been described by Keith *et al.*[124] (phenols), by Kaufmann and Kessen[125] (higher aliphatic alcohols, useful for the study of waxes), by Cerutti and Wezzini[126] (polyalcohols and glycols), and by Vasyunina *et al.*[127] (polyalcohols in mixtures).

The general principle is that the different spots are cut out, eluted with a suitable solvent, and the absorbance of the eluate is then determined and compared with a calibration curve.

### (*b*) *Thin-layer Chromatography*

Here, too, many of the methods recommended for qualitative separation of phenols, alcohols, and their derivatives (see Part B, pp. 50–55) allow a semiquantitative or quantitative microdetermination.

Particularly useful are, according to Gasparič and Gemzova,(128) the *p*-hydroxyazo dyes obtained by coupling the phenol with diazotized *p*-nitro-aniline or sulphanilic acid. Acetylation of the hydroxyazo dyes results in characteristic changes in the chromatographic behaviour of the compound under study. These changes can be expressed quantitatively by means of a term $R_M$, introduced by Bate-Smith and Westall,(129) and defined by the expression: $R_M = \log[(1-R_f)-1]$. Gasparič and Gemzova claim that $\Delta R_M$-values defined as $\Delta R_M = R_{M''} - R_{M'}$, where $R_{M'}$ is the $R_M$-value of the hydroxy compound and $R_{M''}$ the $R_M$-value of the acetylated compound, are obtainable with satisfactory reproducibility, indicating not only the presence of one or more hydroxy groups, but also their number and their type.

### (*c*) *Gas Chromatography*

Gas chromatographic methods for the identification of nearly all types of organic compounds through retention times on polar or non-polar columns have been developed during the past 10–15 years. Frequently, the area of the peaks obtained permits a quantitative estimation of the compounds in a mixture, thus facilitating the analysis of mixtures to a very considerable degree. Either free hydroxy compounds or suitable derivatives can be used for their qualitative characterization and quantitative estimation by gas chromatography.

(i) *Free aliphatic and alicyclic compounds.* Kuffner and Kallina(130, 131) separated saturated $C_5$ and $C_8$ alcohols on polyethylene glycol columns of 182 cm length, i.d. 8 mm, containing solid phase celite 545 coated with Carbowax 300 from which components with b.p. below 200°C at 2 Torr were removed. The column temperature was 100°C for $C_5$ alcohols, or 163°C for $C_8$ alcohols; the carrier-gas was nitrogen at a rate of 3 l. h$^{-1}$ for $C_5$ alcohols, 1 l. h$^{-1}$ for $C_8$ alcohols. They also tried columns of castor oil or lanolin, but the best results were obtained with the polyethylene glycol columns.

TABLE 10.1. ABBREVIATIONS FOR SUPPORTS AND COATINGS USED IN GAS CHROMATOGRAPHY

| | |
|---|---|
| Alkaterge | Amine surfactant |
| Apiezon | Polymer of hydrocarbons |
| Carbowax | Polyglycols |
| Chromosorb | Polymer of hydrocarbons |
| DEGA | Poly-diethylene glycol adipate (cross-linked with pentaerythritol) |
| DEGS | Diethylene glycol succinate |
| Diatoport | Diatomite support |
| DMCS | Treated silicone grease on chromosorb |
| EGSS-X | Ethylene glycol succinate-methyl silicone polymer |
| Gas-Chrom | Firebrick or diatomite support |
| NPGA | Neopentyl glycol adipate |
| PEG | Polyethylene glycol |
| PEGA | Polyethylene glycol adipate |
| Porapak | Ethylvinylbenzene-divinylbenzene polymer |
| Reoplex | Polyesters |
| Silicone GE-XE60 | Silicone gum with 50% nitrile |
| Tergitol | Nonylphenyl ether of polyethylene glycol |
| Triton X-305 | Alkyl aryl polyether alcohol |
| Ucon | Polyalkylene glycol fluids |

Other abbreviations can be found in commercial catalogues, e.g. from Analabs, Applied Sciences Laboratories, Perkin-Elmer, and others.

Sokolov and Kolesnikova[132] separated $C_1$–$C_4$ alcohols on a column of firebrick, 40–50 mesh, calcinated at 550–600°C, cooled and treated with a 30% solution of cetyl alcohol in ether (the latter was then evaporated). The column was filled under helium pressure and then heated for 1·5–2 h at 50–100°C. The sample was introduced by means of a dosimeter. The carrier gas was helium at a rate of 160 ml $min^{-1}$; the column and detector temperature = 90°C.

Tritosyl phosphate was also tried as a column coating, but the separation of the alcohols was unsatisfactory.

Hofmann and Struppe[133] separated $C_1$–$C_9$ alcohols on a 2·7-m column (diameter 3 mm) of Sterchamol coated with 10% Reoplex, at temperature 160°C, with carrier-gas hydrogen at 4·5 l. $h^{-1}$, or better on a 6 m × 3 mm column of Sterchamol coated with 25% polyethylene glycol 2000, at temperature 200°C, with carrier-gas hydrogen at 6 l. $h^{-1}$. $C_{10}$–$C_{20}$ alcohols were separated on a 1·3 m × 3 mm column of Sterchamol coated with 15% Reoplex, at temperature 210°C, with carrier-gas hydrogen at 4·5 l. $h^{-1}$.

Robinson[134] separated alcohols on silicone-coated columns. For alcohols up to $C_{13}$ a 3-m column, for higher alcohols a 1-m column, was used in order to keep the retention times below 30 min. Tetradecanol (10–15%) was used as internal standard up to $C_{13}$ alcohols, pentanol (30–40%) for higher alcohols. The inlet temp was 275°C, programmed temperature range 50–300°C, heating rate 7–9°C min$^{-1}$, detector temp 275°C, carrier-gas helium, 55 ml min$^{-1}$ for $C_{13}$–$C_{20}$ alcohols, 35 ml min$^{-1}$ for $C_5$–$C_{13}$ alcohols. Branched chain alcohols showed shorter retention times than straight chain alcohols.

Rudol'fi *et al.*[135] separated the isomeric menthols on a column coated with ethylene glycol adipate, at column temp 130°C, with carrier-gas hydrogen 90 ml min$^{-1}$.

In many instances alcohols are prepared by a hydroboration reaction. Ikeda *et al.*[136] pointed out that a gas chromatographic examination of the reaction mixture can be invalidated by removal of alkyl borates if boric acid has not been completely removed before subjecting the sample to the examination.

Schlunegger[137] determined $C_1$–$C_5$ alcohols in aqueous solution by gas chromatography on steel columns, 12 ft × $\frac{3}{4}$ in. i.d., solid phase Chromosorb A 60–80 mesh coated with 10% NGPA and 0·2% Alkaterge T. A double flame ionization detector with carrier-gas nitrogen at 17 ml min$^{-1}$ was used. Before use the column was heated for 24 h at 140°C. Programmed temperature rise: 6 min isothermal at 70°C, then temperature rise 1°C min$^{-1}$ to 100°C.

For the separation of water and $C_1$–$C_5$ alcohols Bluestein and Posmanter[138] preferred a 12 ft × $\frac{1}{4}$ in. column packed with 5% Triton X-305 on teflon, using helium as carrier-gas and a thermal conductivity detector. The lower limit of detection of any component was about 0·3% of the mixture. Even at high water concentration the peaks showed only negligible tailing. The calculation of unknown samples was made by a new variation of the internal normalization method which is applicable to any three or more component systems.

Doelle[139] separated $C_1$–$C_7$ alcohols on a column of celite (60–80 mesh) coated with 30% (w/w) polyethylene glycol 400, hydrogen flame detector, column temp 70°C up to $C_4$ alcohols, 110°C for $C_5$–$C_7$ alcohols.

Schwartz and Mathews[140] separated $C_1$–$C_5$ alcohols by capillary column gas chromatography on 200-ft columns, 0·0625 in. o.d., 0·010 in. i.d. with a 15-μ finish coated with a 9/1 blend of Igepal CO-210 (nonylphenol with an average of 1·5 mole of ethylene oxide) and Igepal CO-530 (nonylphenol with an average of 6 moles of ethylene oxide), resulting in a liquid phase composed of nonylphenol with 1·95 mole of ethylene oxide. The alcohols were introduced as 10% (v/v) solutions in chloroform. Temp 65°C, pressure 15 lb in$^{-2}$ hydrogen. Detector: hydrogen flame ionization.

Minyard *et al.*(141) combined gas chromatographic separation on Carbowax 4000 columns coated with Apiezon L with derivatization of the separated alcohols by reaction with 3,5-dinitrobenzoyl chloride or *o*-nitrophenylisocyanate (see Part B, 5.4, p. 30 and 7.2 *b* (ii), p. 47). A 20-gauge hypodermic needle 1 cm long was attached to the Luer-Lock GLC exhaust port to cool the emerging compounds before leading them to the reagent solution (1 g of 3,5-dinitrobenzoyl chloride in 7·5 ml *p*-xylene and 1·0 ml tetrahydrofuran or 10 g *o*-nitrophenylisocyanate in 100 ml benzene, redistilled over calcium oxide). All primary terpene alcohols examined, however, reacted better with 3,5-dinitrobenzoyl chloride at an exhaust temp of 185°C with the needle off, but secondary alcohols gave better yields at 155°C with the needle on. With *o*-nitrophenylisocyanate, all types reacted better at 185°C with the needle off.

Kolesnikova *et al.*(142) succeeded in separating isomeric alkyl-substituted cyclohexanols gas chromatographically, using triethanolamine as the stationary phase. The alcohols could be separated from the corresponding ketones using the same stationary phase. Not only structurally isomeric but also stereoisomeric alcohols could be separated.

Kotel'nikov and Nichikova(143) separated $C_7$–$C_{17}$ primary alcohols on a celite 545 (100–120 mesh) column coated with a mixture of 10% silicone rubber E-301 and 4% polyethylene glycol 20.000 at column temp 160°C.

For secondary alcohols a column of celite 545 (100–120 mesh) coated with 10% polyethylene glycol adipate at column temp 180°C and 160°C was used.

(ii) *Glycols.* Balakhontseva and Poltinina(144) determined glycols gas chromatographically on a 2-m column, stationary phase diatomite quartz coated with polyethylene glycol 10.000, column temp 138–145°C, carrier-gas helium at 60–70 ml min$^{-1}$.

Weatherall(145) described the preparation of columns suitable for the separation of $C_2$–$C_6$ aliphatic diols and some alicyclic diols. Columns: Celite (60–80 mesh, 250 g) was added to petroleum ether b.p. 60–80°C (1·2 l.) containing hexamethyldisilazane (62·5 ml) and the mixture was refluxed for 20 h with exclusion of atmospheric moisture. The solid was washed with petroleum ether b.p. 60–80°C (2 × 500 ml), propanol (500 ml), again with petroleum ether b.p. 60–80°C (2 × 500 ml) and finally with petroleum ether b.p. 30–40°C (2 × 500 ml). It was then transferred to a wide-necked conical flask (2 l) and dried on a steam-bath for 2 h in an atmosphere of nitrogen. To dry silanized celite (158 g) was added chloroform (750 ml), then, with vigorous swirling by hand, 27·4 g (15%) of LAC-2R-446 (diethylene glycol adipate cross-linked with pentaerythritol) in chloroform (150 ml) and tergitol (0·6 g). Swirling was continued for 5 min and the solvent was removed on a rotatory evaporator until the solid was free-flowing.

The above packing was used to prepare 20 ft × $\frac{3}{4}$ in. (o.d.), 25 ft × $\frac{3}{16}$ in. (o.d.) and 6 ft × $\frac{1}{4}$ in. (o.d.) columns. In all cases copper tubing was used, the tubing being held vertically and filled slowly under gravity with manual tapping. After coiling and before use the columns were conditioned at 170°C for 20 h in a stream of nitrogen.

Lindsay Smith and Waddington[(146)] studied the efficiency of different column fillings for the separation of aliphatic alcohols and diols. The materials studied were PAR-1, PORAPAK Q, PAR-1 + PEG 1500, celite + Apiezon L, and celite + PEG 1500.

Novoselov *et al.*[(147)] separated $C_2$–$C_4$ diols and estimated their content in dilute alcoholic or aqueous solutions, using 2-m columns, solid phase INZ-600, TND-SM, Sperochrom-1 or Sterchamol coated with 12–15% polyethylene glycol 600 or 2000. Column temp 105–130°C; carrier-gas nitrogen; 80–100 ml min$^{-1}$.

(iii) *Phenols*. Bergman and Jentzsch[(148)] estimated the contents of phenol mixtures by gas chromatography on 2-m columns filled with celite coated with dicetyl phthalate (1-m columns) or silicone oil (Dow Corning 2-m columns). Maximum column temp 150°C; carrier-gas helium at 6–14 l h$^{-1}$, depending on the composition of the mixture. The separation of methyl- and ethylanisoles was also studied.

Janák *et al.*[(149–151)] studied the separation and estimation of monohydric and dihydric phenols on columns of celite coated with Apiezon L or Apiezon M (both monohydric and dihydric phenols), erythritol (monohydric phenols), mannitol or dulcitol (dihydric phenols).

Fitzgerald[(152)] determined the gas chromatographic retention times for more than fifty phenols with various stationary phases. He indicated that partial separation could be obtained both with polar and with non-polar stationary phases. In cases where the separation is only partial, complete separation may be obtained if both types of column are used in turn. Apiezon L and sodium dodecylbenzenesulphonate were found to be the best non-polar and polar phases respectively.

A study of the separation and estimation of phenolic carboxylic acids has been reported by Méndez and Stevenson,[153] and of phenols by Méndez.[154] Columns with solid support 60–80 mesh Diatoport-S (diatomaceous earth treated with dichlorodimethylsilane) coated with silicone gum rubber (SE 30) or diethylene glycol succinate (DEGS) used in 5 ft copper columns $\frac{1}{4}$ in. o.d., $\frac{1}{8}$ in. i.d., provided well-defined peaks for a series of phenolic carboxylic acids. The carrier-gas was helium at a rate of 42 ml min$^{-1}$ at the exit port; column temp 100°C for 3 min and then programmed to 250°C with a temp rise of 25°C per min (for di- and trihydric phenols).

Landault and Guiochon[155] described the analysis of phenol, methylphenols, and polymethylphenols resulting from the pyrolysis of phenol-formaldehyde polycondensates, using gas chromatography on packed capillary columns 5 m × 0·5 mm (i.d.) for the estimation. The solid support is 80–100 μ Chromosorb P, washed with aqua regia. Once this column is drawn and the fittings are soldered the support is silanized by percolating a solution of 10% dichlorodimethylsilane in benzene at a rate of 2 m min$^{-1}$ for 15 min. The solution is then flushed away with dry nitrogen and the column heated at 60°C overnight with a slight gas current. A 1% solution of a mixture of di-(3,3,4-trimethylcyclohexyl) *o*-phthalate and tri-2,4-xylenyl phosphate in pure, freshly distilled methylene chloride is then percolated at the same rate for 15 min and the column is again dried as indicated.

The chromatographic separation is carried out with hydrogen as carrier-gas and flame ionization detector.

The carrier-gas flow-rate cannot be measured on the working column but its velocity is easily derived with precision from the emergence of methane, which is not retained, and from the decompression factor, *j*. The carrier-gas velocity is the physical parameter of interest, and the authors therefore quote the optimal velocity in each case studied.

Clifford and Watkins[156] studied the separation of dinitroalkyl phenols on a 6 ft × $\frac{1}{4}$ in. (o.d.) copper tube column packed with 5% DEGA + 0·4% 85% phosphoric acid on acid-washed Chromosorb W (60–80 mesh), carrier-gas helium at 60 ml min$^{-1}$, working temp 215°C, injector temp 300°C and detector temp 275°C.

Malz and Gorlas[157] described a method for analysing the phenols in gas condensates. The phenols were absorbed in sodium hydroxide. 5 g of the phenate solution was acidified with 3 ml dilute sulphuric acid, and sufficient silica gel to absorb all the liquid was added. The silica gel layer was then placed on a silica gel column, 70 cm long, 10 mm i.d., with a water content of 15–20%. 70 ml of cyclohexane was poured onto the column and all the neutral oils were contained in the first 50 ml of eluent. Then the column was percolated with isopropyl ether and this eluted all phenols. The eluate was collected in an Erlenmeyer flask (100 ml) containing a weighed quantity of durol as internal standard, and gas chromatographed on one of the standard columns for separation of phenols.

Delaunois and Bettens[(158)] described the complete separation and analysis of phenolic constituents from the thermal cracking of phenols, using a 100-m capillary column loaded with propylene glycol and run at precisely 150°C. Compared with a standard 2-m column the capillary column was slower (2·5 h versus 45 min), but the separation was improved, as were the ease and accuracy of the determination.

Nosal *et al.*[(159)] carried through the gas chromatographic analysis of chlorophenols obtained by direct chlorination of phenols with gaseous chlorine. All possible dichloro- and trichlorophenols were separated and determined on a column filled with 9·1% Carbowax 20 M or 16·7% Ucon 5OB-55OX on Chromosorb WAW with the addition of 1·5% phosphoric acid. Nitrogen, hydrogen, or air at 60–75, 70–78 or 450 ml $min^{-1}$ was used as carrier-gas; column temp 160–169°C.

Zakupra and Dobrov[(160)] compared the isothermal and the linearly programmed separation of monohydric alkylphenols on non-polar (Apiezon L) and weakly polar (Polyphenyl ether) columns; the carrier-gas was helium at 40–45 ml $min^{-1}$. The programmed separation gave the best results, starting with isothermal separation for 12 min at 110°C, then raising the temp 5°C per min to 175°C.

(iv) *Derivatives of hydroxy compounds*

1. *Acyl derivatives.* Sato *et al.*[(161)] tried to separate even-numbered alcohols $C_6$–$C_{22}$ on columns of Chromosorb coated with silicone grease (80 g Chromosorb, 35–60 mesh, treated with aqua regia, dried, and impregnated with 20 g silicone grease dissolved in ether. The ether was removed on a steam-bath and the coated Chromosorb dried in an oven at 110°C). Columns of stainless steel, 2 mm i.d., were prepared in lengths from 0·75 to 3 m. Operating temp from 180°C to 240°C. However, the free alcohols gave unsymmetrical peaks, and Sato *et al.* therefore treated them

with ketene, converting them quantitatively into the corresponding alkyl acetates, and here the peaks were symmetrical, making the quantitative evaluation more reliable.

Larkham and Pagington(162) heated neutral, alcohol-containing mixtures with chloroacetyl chloride and separated the chloroacetate mixture gas chromatographically, using a halogen-sensitive detector. They claimed that the method was rapid and very sensitive with negligible interference from non-alcoholic compounds.

Wood(163) analysed $C_{14}$–$C_{20}$ alcohols as trifluoroacetates either on columns (5 ft × $\frac{1}{8}$ in., stainless steel) packed with 5% methyl silicone polymer (SE 30), coated on 60–80 mesh Chromosorb W or on 5 ft × $\frac{1}{8}$ in. Pyrex columns packed with 15% ethylene glycol succinate-methyl silicone polymer (EGSS-X) coated on 100–110 mesh Gas-Chrom P. The SE 30 column was operated at 190°C, the EGSS-X column at 130°C with helium as carrier-gas and a hydrogen flame ionization detector.

Neurath and Lüttich(164) have studied and separated 102 alcohols as alkyl 4′-nitroazobenzene-4-carboxylates (for preparation, see Part B, 5.5 p. 30) on 1- or 2-m stainless steel columns, 4 mm i.d., packed with 2·5% (w/w) silicone grease on 60–80 mesh Chromosorb G, acid-washed and DMCS treated, using a gas chromatograph with flame ionization detector.

Volodina and Kon'kova(165) found that isopropenyl acetate reacts quantitatively with primary alcohols in 1·5 h and with secondary alcohols in 3 h or more to form stable derivatives:

$$\mathrm{ROH + CH_2{=}\underset{\displaystyle OCOCH_3}{\underset{|}{C}}{-}CH_3 \rightarrow ROCH_2{-}\underset{\displaystyle OCOCH_3}{\underset{|}{CH}}{-}CH_3}$$

These derivatives could be determined gas chromatographically. Heat a mixture of 50–100 mg of the alcohol mixture, 100–200 mg isopropenyl acetate and 1–2 mg *p*-toluenesulphonic acid on a water-bath at 80°C for 3 h. Add 5 μl of the reaction mixture to a column, 2 m × 6 mm, filled with polyethylene glycol 3000 on Chromosorb (60–80 mesh) 1/9 and operate the column at 80–150°C, with helium carrier-gas at 60–70 ml min$^{-1}$.

Galetto *et al.*(166) separated and estimated 3,5-dinitrobenzoates on a column coated with SE 30 silicone fluid at a column temp of 250°C. The effluent material from the peaks was trapped and shown by infrared spectroscopy and thin-layer chromatography to be the original derivatives injected. Mixtures of the derivatives could be separated successfully without previous purification.

2. *Ethers.* Bhattacharyya *et al.*(167) estimated monohydric phenols by quantitative methylation with excess dimethyl sulphate and molecularized sodium in dioxan. The anisole mixture was separated on a celite column coated with Apiezon L.

Cohen *et al.*(168) described the trace determination of phenols by gas chromatography of their 2,4-dinitrophenyl ethers on a Silicone GE XE-60 column (see Cohen and Wheals(169)), making possible the separation of phenols by thin-layer chromatography or paper chromatography, using etherification on the plate or on the paper-sheet, followed by elution of the ether and injection in the gas chromatograph. For details, see recommended Method *R*, p. 141.

The trimethylsilyl ethers of both aliphatic, alicyclic and aromatic hydroxy compounds have been found to be extremely useful for the gas chromatographic separation and estimation of hydroxy compounds. The preparation of the ethers has been described in Part B, 6.1 (*c*), p. 38. Some examples of their use for estimations are given here.

Langer *et al.*(170) preferred the trimethylsilyl ethers of phenols to the methyl ethers for the gas chromatographic separation, using a 3·66 m × 3 mm column packed with 34 g firebrick (42–48 mesh) coated with 12% (w/w) of dibutyl tetrachlorophthalate or Dow Corning Silicone 550. Working temp 125°C, carrier-gas helium at 153 ml $min^{-1}$.

Freedman and Croitoru(171) described the quantitative estimation of even hindered phenols by complete trimethylsilylation (see Part B, recommended Method *Q*, p. 62). The ethers were then separated on a copper capillary column of 150 ft, i.d. 0·010 in., coated with dioctyl sebacate. Working temp 125°C, carrier-gas helium at 2 ml $min^{-1}$, measured at 125°C.

Sweeley *et al.*(172) trimethylsilylated polyols by treating 10 mg of substance with 1 ml anhydrous pyridine (kept over potassium hydroxide pellets), 0·2 ml hexamethyldisilazane and 0·1 ml chlorotrimethylsilane in a 1-dram plastic-stoppered vial or similar container. Shake the mixture vigorously for about 30 sec and allow to stand for 5 min or longer at room temp. The solution becomes cloudy (separation of ammonium chloride) but this does not interfere with the subsequent gas chromatography on a column of SE 52. Working temp 125°C for $C_2$–$C_3$ polyols and 140°C for $C_4$–$C_7$ polyols.

Leibman and Ortiz(173) estimated the trimethylsilyl ethers of glycols on a 4-ft glass column using helium carrier-gas at 100 ml $min^{-1}$.

Dutton *et al.*[174] separated and determined quantitatively trimethylsilyl ethers of polyols (including some carbohydrates) on dual 8 ft × 0·25 in. copper columns containing 20% SF 96 on 60–80 mesh Diatoport S, using a thermal conductivity detector. The columns were heated isothermally at 130°C for 6 min, then programmed at 3°C per min to hold at 220°C. Temperature of injector port 270°C, and of detector block 295°C. The carrier-gas was helium at flow-rate 88 ml $min^{-1}$.

van Ling[175] described the trimethylsilylation and gas chromatographic analysis of aqueous solutions of polyhydric alcohol mixtures, the silylating reagent being Tri-sil "Z" (Pierce Chem. Co.), a mixture of trimethylsilyl-imidazole in dry pyridine, 1·5 m-equiv $ml^{-1}$.

Introduce 10–15 mg of sample and 1 ml Tri-sil "Z" into a stoppered vial. Shake well, and allow to stand for several min. The temp may be raised to 60–70°C if necessary to dissolve the sample. Usually the silylation is complete within a few min. Inject 25 μl into the gas chromatograph, having a 1·5-m stainless steel column 4 mm i.d., packed with 20 wt.% Silicone rubber U.C.W. 98 on 60–80 mesh Diatoport S. The column is conditioned with Silyl 8 (Pierce Chem. Co.). Thermal conductivity detector. Carrier-gas hydrogen at 60 ml $min^{-1}$.

3. *Azo compounds*. Kushner *et al.*[176] described an improved method for the estimation of volatile phenols. The mixture is separated by thin-layer chromatography on Silica gel G/Cellulose MN plates, the plates being dried at room temp for at least 6 h, and developed with cyclohexane/ethyl acetate (5/1 v/v). A reference plate is sprayed with diazotized *p*-nitroaniline followed by 10% aqueous sodium carbonate to localize the spots. By superimposing this plate over the other it is possible to scratch the individual spots off the plate. They are transferred to a microfilter funnel with a sintered-glass disc (porosity 0·06–0·08 μ), washed consecutively with 4 ml and 14 ml of ethyl acetate/cyclohexane (1/1). The first 4 ml of wash solution, containing about 70% of the phenols, is set aside. The second extraction is reduced in volume by a stream of nitrogen to about 1 ml, then both extracts are combined and further reduced to a final volume of 2 ml.

Samples of the extracts are injected into a gas chromatograph with a 6 ft × $\frac{1}{8}$ in. stainless steel column packed with 10% U.C.W. 98 silicone on 80–100 mesh Chromosorb W. Programmed temp rise 1°C $min^{-1}$, flame ionization detector, with detector and injector port at 195°C and 250°C, respectively.

## 14. Mass Spectrometry

Zahn *et al.*[177] found it impossible to apply mass spectrometry to the determination of alcohols in the presence of hydrocarbons because of the similarity of their mass spectra, but when the alcohols were first converted into trimethylsilyl ethers, characteristic mass spectra were obtained free of interference from hydrocarbon ion fragmentation. Not only is mass spectrometric determination of alcohols now possible without prior separation from other materials, but the range of a mass spectrometer at room-temperature is also extended to include the $C_{10}$-alcohol region. Water does not interfere because it is converted into hexamethyldisiloxane which has its own distinctive mass-spectral peak.

## 15. NMR Spectrometry

Mathias[178] found it possible to estimate the relative amount of primary and secondary hydroxy groups using nuclear magnetic resonance spectra. He recommended acetylation of the alcohols before determining the spectra, because there is a greater difference between the δ-values of $—COOCH_2R$ and $—COOC(R)(R')—H$ than between the δ-values of $RCH_2OH$ and $CH(R)(R')OH$. Thus, both the detection and the estimation of the two types of hydroxy function are facilitated.

Chapman and King[179] found that dimethylsulphoxide was the best solvent for the classification of alcohols by NMR spectra because strong hydrogen bonding to the solvent shifts the hydroxyl resonance downfield and reduces the rate of proton exchange sufficiently to permit observation of hydroxyl proton splitting. It is possible to use dimethylsulphoxide instead of dimethylsulphoxide-$d_6$.

The hydroxyl proton in methanol gives a quartet; primary, secondary, and tertiary alcohols give clearly resolved triplets, doublets and singlets, respectively.

Polyhydroxy compounds give separate peaks for each hydroxy proton.

Addition of a few drops of deuterium oxide is sufficient to remove the hydroxyl hydrogens.

Godsey[(180)] studied the NMR spectra of structural isomers of secondary straight-chain alcohols and their 3,5-dinitrobenzoates. He found that the chemical shifts are correlated (empirically) with the number of C—C bonds between the carbon of the methyl group and the carbon adjacent to the functional group. Correlation data from the 3,5-dinitrobenzoates may be used to characterize the 2- through 6-position isomers of the secondary alcohols.

For estimation, authentic spectra of the alcohols should be compared with the spectrum of the sample.

Babiec *et al.*[(181)] recommended the preparation of dichloroacetates of primary, secondary, and tertiary alcohols for the characterization through NMR spectroscopy, because the dichloroacetate proton resonance is removed from most alkyl and aryl resonances and has the advantage of giving an extremely sharp resonance line. The latter fact is particularly advantageous when mixtures of dichloroacetates are studied. They found distinct separation between the dichloroacetyl resonance of esters of tertiary alcohols on the one hand and primary and secondary alcohols on the other hand. This separation can be enhanced by suitable choice of solvent. Generally, the order of shielding is tertiary > secondary > primary, resulting in the tertiary resonance falling to higher fields and the primary resonance to the lowest field. However, there is no clear separation of primary and secondary resonances.

Dimethylsulphoxide-$d_6$ is preferable to deuterochloroform as it enhances the chemical shift separation of the individual dichloroacetyl resonances as well as the tertiary, secondary, and primary group separation.

CHAPTER 11

# RECOMMENDED METHODS FOR THE ESTIMATION OF HYDROXY COMPOUNDS

## 1. Titrimetric Methods

### *Method A. Visual Titration of Phenols (Fritz and Keen[(6)])*

(*a*) *Negatively substituted phenols.* Add 2 drops of a saturated solution of azoviolet in benzene to 20 ml dimethylformamide and neutralize acidic impurities by titration with 0·1 N potassium methoxide in methanol/benzene (clear blue colour). Add the sample of phenol and titrate again to the clear blue colour. Standardize the titrant against benzoic acid.

When *a* g of the phenol require *b* ml of base of normality *n* the equivalent weight, *E*, of the phenol is given by:

$$E = \frac{a \cdot 1000}{b \cdot n}$$

(*b*) *Phenol and alkylsubstituted phenols.* Add 2 drops of 0·15% solution of *o*-nitroaniline in benzene to 25 ml ethylenediamine. Titrate with 0·1 N potassium methoxide in benzene/methanol until the colour changes from clear yellow to orange-red. Add rapidly the sample of phenol and titrate again to the orange-red end-point.

The equivalent weight, *E*, is calculated as above.

It is essential that the methanol concentration in the titrant is as low as possible. The following procedure for the preparation of the titrant is recommended.

Dissolve 4 g potassium in 40 ml methanol and 50 ml benzene. Add methanol to the cloudy solution until it becomes clear; benzene is then added, slowly and with continual stirring or shaking, until the mixture again becomes cloudy. Alternate additions of methanol and benzene are then continued until the solution has been diluted to 1 l.

(*c*) *Phenol carboxylic acids.* In both procedures (*a*) and (*b*) carboxylic groups, if present, are titrated together with the phenol groups. A differentiation titration in which only the carboxylic groups are titrated can be carried out by dissolving the substance in 30–40 ml acetone or acetonitrile and titrating the solution (indicator: 2 drops of 0·2% *p*-hydroxyazobenzene) to a clear yellow end-point. From the difference in volume of titrant used up in procedure (*a*) or (*b*) and in procedure (*c*) the equivalent weight of the phenol can be calculated.

### *Method B. Potentiometric Titration of Phenols (Morman and Harlow*[8]*)*

The system glass electrode/calomel electrode is used for the titration. The calomel electrode is connected to the titration solution by a non-aqueous bridge of 3-methylsulpholan saturated with tetraethylammonium perchlorate in order to avoid introducing water and potassium ions both of which would depress the potential span. The saturated potassium chloride solution in the calomel electrode is replaced with 0·1 N solution of tetramethylammonium chloride to avoid precipitation of potassium perchlorate in the ground-glass junction.

Dissolve the sample in sulpholan, 3-methylsulpholan, or 2,4-dimethylsulpholan. Sulpholan has m.p. 28°C but addition of the sample and the first increments of titrant lowers the m.p. sufficiently to avoid freezing. Titrate with 0·2 N tetrabutylammonium hydroxide in propan-2-ol (Harlow and Wyld[11]).

Phenol shows two inflections due to association through hydrogen bonding:

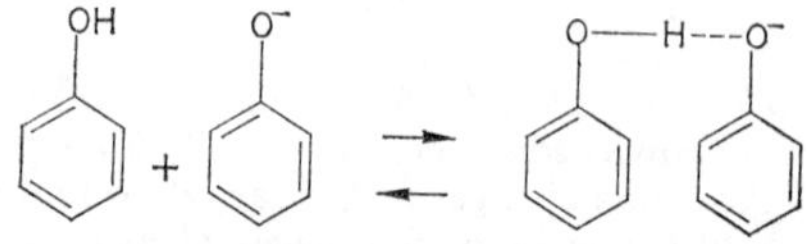

(cf. van der Heijde[2] and Harlow and Bruss[12]), but 2,6-di-*tert*.-butyl-4-methylphenol and other hindered phenols show only one.

The procedure is excellently suited to differential titration of acids of different strength, due to the unusually wide potential span.

The equivalent weight, *E*, of the substance titrated is calculated as above.

### *Method C. Thermometric Estimation of Phenolic Hydroxyl Groups (Vaughan and Swithenbank*[21, 22]*)*

Coal is ground to pass through a 72-mesh B.S. sieve (less than 210 μ). About 30 mg are transferred to the agate capsule of a high-speed vibration mill, 100 Hz, with agate capsule of internal dimensions of about 30 mm length and 10 mm diameter with agate ground-in stopper and two agate balls; the agate balls are added and the capsule closed with the stopper. Attach the capsule to the vibration arm of the mill and adjust the position of the capsule to obtain maximum vibration. Grind the sample for 5 min, then add

0·5 ml pyridine to the capsule and grind for a further hour. Unfasten the capsule from the mill, wash its contents with acetone into the titration vessel through a small funnel to remove the agate balls, until 5 ml have been collected in the titration vessel as shown by the mark. Assemble the vessel on the thermometric titration apparatus and titrate with N potassium hydroxide in propan-2-ol at a rate of 0·1 ml $min^{-1}$ until a straight-line portion of the temp rise has been recorded.

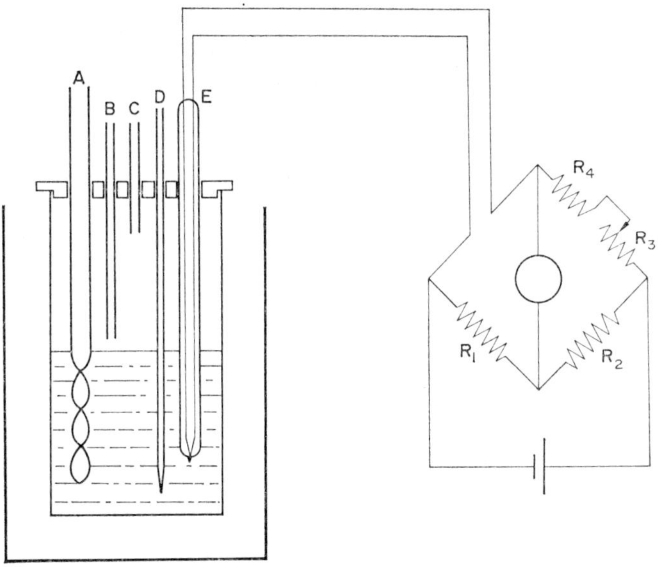


FIG. 7. Apparatus for thermometric titration of phenols (Vaughan and Swithenbank[21]).

| | |
|---|---|
| A. Glass stirrer | $R_1$ 1000-ohm resistor |
| B. Acetone inlet | $R_2$ 1000-ohm resistor |
| C. Air inlet | $R_3$ 1000-ohm resistor |
| D. Titrant tubing | $R_4$ 1500-ohm resistor |
| E. Thermistor | |

(Reprinted by permission of the Society for Analytical Chemistry and the authors.)

Measure the chart length from the start to the tangential end-point (shown as a dotted line in Fig. 8) and subtract the blank value. To ascertain the blank value introduce by pipette 0·5 ml pyridine into the titration vessel and dilute to the 5-ml mark with acetone; titrate the mixture.

To standardize the titrant, weigh about 7 mg benzoic acid into the titration vessel, add to it by pipette 0·5 ml pyridine, make up to the 5-ml mark with acetone and titrate as above. Subtract the blank value, and calculate the weight in milligrams of OH corresponding to 1 mm of the recorder chart.

### *Method D. Titration with Lithium Aluminium Dibutylamide* (*Small*,[24] *Jordan*[25])

*Reagent:* Lithium aluminium hydride (95%), 1,2-dimethoxyethane, tetrahydrofuran, dibutylamine. Indicator: 0·05–0·1% solution of N-phenyl-*p*-aminoazobenzene in benzene.

*Procedure:* Place 750 ml 1,2-dimethoxyethane (dried over molecular sieve 4A or 5A or freshly distilled from lithium aluminium hydride) in a 1000-ml 2-necked flask fitted with ground-glass joint and vertical condenser. Add 2–3 drops of indicator solution and just enough lithium aluminium hydride to change the colour of the solution from yellow

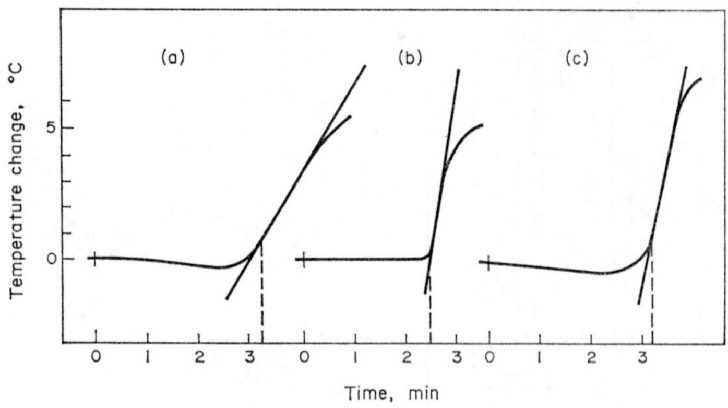


FIG. 8. Thermometric titration curves (Vaughan and Swithenbank[22]). (a) Coal, (b Phenol, (c) 2,6-di-*tert*.-Butyl-4-methylphenol. (Reprinted by permission of the Society for Analytical Chemistry and the authors.)

to red; then add an additional 1·5 g lithium aluminium hydride to the solution and reflux for 25 min. Add, in increments, 31 ml dibutylamine slowly through the condenser, allowing the reaction for hydrogen gas formation and evolution to subside between increments. Continue to reflux for an additional 10–15 min and then remove the condenser, blanket the reagent with nitrogen, stopper, and allow to cool. When cool, transfer it under nitrogen cover to the reservoir of the burette and keep it under slight pressure of nitrogen.

Place 200 ml freshly distilled tetrahydrofuran in a 3-necked flask equipped with a stopper to admit a stream of nitrogen, another stopper to admit the burette tip loosely, and a teflon-covered magnetic stirrer. Add a few drops of indicator solution, start the magnetic stirrer and titrate to the end-point (colour change from yellow to red). Weigh the sample into the flask (from a weighing pipette or other convenient device), allow time for mixing or dissolution, and again titrate to the end-point. Standardize the titrant by titrating a sample of butyl alcohol.

The reactions are:

$$LiAlH_4 + 4(C_4H_9)_2NH \rightarrow LiAl[N(C_4H_9)_2]_4 + 4H_2$$

$$LiAl[N(C_4H_9)_2]_4 + 4ROH \rightarrow LiAl(OR)_4 + 4(C_4H_9)_2NH$$

It is seen that 1 mole of alcohol will use up 1 equivalent of the reagent. The equivalent weight of the hydroxy compound is therefore calculated as in Method *A*, p. 113.

## 2. Acylation Methods

*Method E*

(*a*) *Acetylation with acetic anhydride, catalyzed by perchloric acid* (*Fritz and Schenk,*[27] *Schenk and Fritz*[28]).

*Reagents:* 2 M acetic anhydride in ethyl acetate containing perchloric acid. Dissolve 4 g (2·35 ml) of 72% perchloric acid in 150 ml ethyl acetate in a 250-ml glass-stoppered flask. Pipet 8 ml acetic anhydride into the flask and allow the mixture to stand at room temp for 30 min. Cool to 5°C, add 42 ml acetic anhydride, keep the mixture at 5°C for 1 h and then allow it to warm up to room temp. The solution takes on a faint yellow colour but this has been found not to interfere with the determination of the end-point.

0·25 M acetic anhydride in ethyl acetate containing perchloric acid. Dilute one volume of the above solution with seven volumes of ethyl acetate.

0·55 M sodium hydroxide in water/methyl cellosolve. Dilute 185 ml of saturated aqueous (carbonate-free) sodium hydroxide solution with 430 ml water and then stir it into 5400 ml methyl cellosolve. Standardize the solution against potassium hydrogen phthalate.

0·15 M sodium hydroxide. Dilute 300 ml of the above solution with 800 ml methyl cellosolve containing 10% of water. Standardize against potassium hydrogen phthalate.

*Indicator:* Mix 1 part of 0·1% neutralized aqueous cresol red with 3 parts of 0·1% neutralized thymol blue.

*Procedure:* Weigh exactly 3–4 m-moles of a monohydroxy compound (proportionally less of a polyhydroxy compound) into a 125-ml glass-stoppered flask. Pipet exactly 5 ml of the 2 M acetic acid reagent into the flask and mix the contents by swirling. Leave the flask to stand at room temp for 5 min, then add 1–2 ml water and shake the mixture. Add 10 ml of pyridine/water (3/1) solution and leave the flask to stand for a further 5 min. Add indicator and titrate the solution with 0·55 M sodium hydroxide solution; the end-point is indicated by a colour-change from yellow to violet. Dark-coloured solutions are better titrated potentiometrically to pH 9·8, using a glass electrode/calomel electrode assembly.

Run a blank in parallel by pipetting 5 ml of the reagent into a 125-ml flask containing 1–2 ml water and proceeding as indicated above after addition of water.

When $a$ g of the substance requires $b$ ml sodium hydroxide of normality $n$, and $b_1$ ml sodium hydroxide are used up in the blank, the equivalent weight, $E$, of the hydroxy compound is given by:

$$E = \frac{a \cdot 1000}{(b_1 - b) \cdot n}$$

**N.B. Organic solutions containing hydroxy compounds and perchloric acid should never be heated, nor should they be left to stand, as serious explosions may occur. Such solutions should be disposed of immediately after the determination is complete.**

When only 0·3–0·4 m-equiv of the hydroxy compound are available use 0·25 M acetic anhydride and 0·15 M sodium hydroxide and proceed as indicated above.

Klouček[29] recommends the following procedure for semimicro estimation:

Weigh the sample (15–20 mg) in a 100-ml glass-stoppered flask, add 1 ml of the 2 M acetylating mixture and swirl. Leave the flask for 10 min at room temp, then add 1 ml water and 5 ml pyridine/water (3/1) mixture. Swirl and leave the flask for 5 min, then rinse the stopper with ethanol, add a further 10 ml ethanol, 7 drops of *o*-cresol red and 14 drops of thymol blue (both 0·1 % in 50 % ethanol/water). Titrate with 0·25 N ethanolic sodium hydroxide to the colour-change from blue to violet. Compare with a blank.

Magnuson and Cerri[30] preferred 1,2-dichloroethane as solvent instead of ethyl acetate, an advantage being that no cooling is necessary in preparing the reagent (1 M solution of acetic anhydride in 1,2-dichloroethane, 0·15 N in perchloric acid).

Add to 420 ml 1,2-dichloroethane in a 500-ml glass-stoppered flask 6·2 ml 72 % perchloric acid and then, slowly and with stirring, 55 ml acetic anhydride. The reagent is stable for at least 2 months.

*Procedure:* Pipet 10 ml reagent into a 125-ml glass-stoppered flask and add an accurately weighed sample (4–5 m-equiv) of the hydroxy compound. Leave the flask for 5 min, then add 35–40 ml of dimethylformamide/pyridine/water (6/3/1) hydrolyzing solution. Hydrolyze for 10–15 min, then add 5 drops of 1 % thymol blue as indicator and titrate with 0·55 N ethanolic potassium hydroxide to the blue end-point. Dimethylformamide aids in increasing the solubility both of 1,2-dichloroethane and of water, and its presence also sharpens the indicator end-point.

Run a blank in parallel.

*(b) Acetylation with acetic anhydride in ethyl acetate, catalyzed by p-toluenesulphonic acid (Madeline and Debourg,[32] Stetzler and Smullin[33]).*

*Reagent:* 2 M acetic anhydride in ethyl acetate, 0·15 M in *p*-toluenesulphonic acid.

Add 14·4 g *p*-toluenesulphonic acid (with 1 mole of water of crystallization) to 360 ml ethyl acetate in a 500-ml flask. Agitate the mixture with a magnetic stirrer until the acid has completely dissolved. Add, slowly, 120 ml acetic anhydride, maintaining agitation during the addition. A slight yellow colour will develop but the anhydride content of the reagent will remain at a satisfactory level for several days.

0·56 M potassium hydroxide in absolute methanol.

*Indicator:* 1 part of 0·1 % neutralized cresol red mixed with 3 parts of 1 % neutralized thymol blue.

*Procedure:* Weigh accurately a sample containing 5–6 m-equiv of hydroxyl into a 250-ml Erlenmeyer flask. Pipet 5 ml of the acetylating reagent into the flask. Cover the flask with a watch glass and immerse it just above the liquid level in a water bath at 50 ± 1°C. After 5 min remove the flask from the bath and swirl gently. Return the flask to the bath and allow the reaction to proceed for a further 10 min. Add 2 ml distilled water and swirl the mixture vigorously. Add 10 ml pyridine/water (3/1) solution, rinsing the sides of the flask during the addition. Allow the flask to stand for 5 min at room temp to hydrolyze the excess acetic anhydride completely. Add indicator and titrate with 0·56 M potassium hydroxide until the colour changes from yellow to violet. Run a blank in parallel.

Pietrzyk and Belisle[34] prefer 2,4-dinitrobenzenesulphonic acid to *p*-toluenesulphonic acid as catalyst for the acetylation.

*Acetylating reagent:* Weigh 6·8 g 2,4-dinitrobenzenesulphonic acid into a 100-ml measuring flask, dissolve it in a small amount of ethyl acetate, add 33 ml acetic anhydride, leave the mixture to stand for 15–30 min, then add ethyl acetate nearly to the mark and adjust with ethyl acetate after 30 min.

*Procedure:* Weigh accurately a sample containing approximately 4 m-equiv of hydroxyl into a 125-ml glass-stoppered flask and add by pipet 3 ml ethyl acetate to dissolve the sample (or pipet into the flask a 3-ml aliquot of an ethyl acetate solution of the sample). Pipet into the solution 5 ml acetylating reagent and allow the reaction to proceed for 5–10 min. Then add 1–2 ml water, swirl the mixture, and add 10 ml pyridine/water (3/1) solution. Leave the mixture to stand for 5 min and titrate with 0·6 N sodium hydroxide. The indicator is an aqueous mixture of 1 part of 0·1% neutralized cresol red and 3 parts of 0·1% neutralized thymol blue; colour change from yellow to violet.

### (*c*) *Acetylation with an acetic anhydride–pyridine mixture* (*Delaby and Sabetay,*[36] *Hauschild et al.*[37]).

*Reagent:* A mixture of 1 part acetic anhydride and 10 parts pyridine.

*Procedure:* Weigh accurately 0·5–1 g of the hydroxy compound into a 250-ml Erlenmeyer flask equipped with a ground-glass reflux condenser and pipet into the flask 100 ml reagent. Heat the flask in a boiling water-bath for 0·5–2 h, taking care that the water-level in the bath is always 2–3 cm above the level inside the flask. Cool, rinse the reflux condenser, the stopper and the walls of the flask with water, then heat the flask for 10 min on the steam-bath to hydrolyze excess acetic anhydride, cool, and titrate with 0·5 N sodium hydroxide, using phenolphthalein indicator. Run a blank in parallel and calculate the equivalent weight of the hydroxy compound from the difference between the two titrations, using the expression given on p. 117.

### (*d*) *Microestimation of primary alcohols* (*Quattrone and Choy*[38]).

*Reagent:* A mixture of 1 part acetic anhydride and 3 parts pyridine.

*Procedure:* Weigh accurately 1–3 mg of the hydroxy compound in a small thin-walled stainless steel boat, 1·25 × 0·35 × 3·5 cm, provided with a glass cover. Add 0·10 ml acetylating reagent (Gilmont microburette) and allow to stand at room temp until the acetylation is complete. Then place the boat at the bottom of a 10-ml titration cell to which 5 ml distilled water is added. Remove the boat cover and add sufficient mercury, by means of a dip pipet, to the boat to fill it. Immerse pencil-type glass- and calomel-electrodes in the solution and place the entire set-up on a magnetic stirrer, the mercury-filled boat acting as a stirring bar. Titrate the excess acetic acid with 1·0 N sodium hydroxide, using a microburette. The end-point is clearly shown at pH about 7·6. Run a blank in parallel.

Most pure primary alcohols are completely acetylated in about 1 h at room temp; ethanol is an exception, requiring a large excess of acetylating reagent. For water–alcohol mixtures 2 h is an adequate time.

For pure alcohols, the equivalent weight is given by the usual expression given on p. 117. For mixtures the percentage of hydroxyl is given by

$$\% \text{ OH} = \frac{b \cdot n \cdot 17 \cdot 100}{1000 \cdot a}$$

when *a* g of the mixture require *b* ml of reagent of normality *n*.

(*e*) *Acetylation with acetyl chloride in toluene* (*Smith and Bryant*[41]).

*Reagent:* 1·5 M acetyl chloride in toluene.

*Procedure:* Pipet 10 ml reagent into a 300-ml Erlenmeyer flask, cooled in ice water, and add 2 ml pyridine. Weigh accurately 4–6 m-equiv (0·2–0·4 g) of the hydroxy compound in a small weighing glass and transfer it to the acetylating mixture. Place the flask in a water-bath at 60°C for 20 min, assisting the reaction by regular shaking of the flask. Then cool the mixture in ice water, add 25 ml water and shake thoroughly to hydrolyze excess acetyl chloride. Add phenolphthalein indicator and titrate with N sodium hydroxide. Run a blank in parallel.

The reactions are:

$$ROH + CH_3COCl \rightarrow CH_3COOR + HCl$$
$$H_2O + CH_3COCl \rightarrow CH_3COOH + HCl$$

10 ml 1·5 M acetyl chloride solution correspond to 30 ml N sodium hydroxide and the maximum difference between the blank and the experiment proper is 50% of the blank. If differences almost equal to the maximum value are obtained, the determination should be repeated with a smaller sample of the substance or with more reagent.

The method may be used for primary and secondary alcohols, polyhydric alcohols, and phenols, but not for tertiary alcohols. Primary and secondary amino groups, mercapto groups and, to some extent, aldehyde groups interfere in the estimation.

(*f*) *Propionylation* (*Pesez*[42]).

*Reagent:* Mix, at least 15 min before use, 30 ml glacial acetic acid containing 1 g *p*-toluenesulphonic acid (not contaminated with sulphuric acid) with 5 ml propionic anhydride.

*Procedure:* Weigh accurately the sample (*a* g) into a glass-stoppered 5-ml flask (0·5–1·5 m-moles of a monohydroxy compound, or proportionally less of a polyhydroxy compound). Add 2 ml reagent, stopper the flask and mix the contents by cautiously shaking the flask, then leave it to stand for 2 h at room temp or heat it on a boiling water-bath for 20 min. Cool, and transfer the flask and its contents to a 500-ml Erlenmeyer flask containing 25 ml of approx 0·1 N aniline in benzene (0·9%) and 30 ml glacial acetic acid. Shake the flask thoroughly and leave it to stand for 5 min at room temp. Add five drops of 0·1% solution of crystal violet in glacial acetic acid, and titrate the solution with 0·1 N solution perchloric acid in glacial acetic acid. Run a blank in parallel.

The reactions are:

$$(C_2H_5CO)_2O + ROH \rightarrow C_2H_5COOR + C_2H_5COOH$$

$$(C_2H_5CO)_2O + C_6H_5NH_2 \rightarrow C_2H_5CONHC_6H_5 + C_2H_5COOH$$

excess aniline being titrated with perchloric acid.

If the consumption of perchloric acid of normality $n$ in the principal experiment is $b$ ml and in the blank $c$ ml, the equivalent weight, $E$, of the hydroxy compound is given by

$$E = \frac{a \cdot 1000}{(b-c) \cdot n}; \quad \% \, OH = \frac{17 \cdot 100}{E}$$

### (g) *Stearoylation* (*Sully*[43]).

*Procedure:* Weigh accurately 12–13 g stearic anhydride into a 250-ml flask equipped with a ground-glass reflux condenser, then weigh accurately the sample of hydroxy compound into the flask. Add 10 ml *m*-xylene and stearoylate by refluxing the mixture for 15–30 min. Cool and add 40 ml pyridine containing 4 ml water, then reflux for a further 15 min to hydrolyze the excess stearic anhydride. Add phenolphthalein and titrate with N sodium hydroxide, the solution being kept hot to avoid crystallization.

Determine in a blank experiment how much N sodium hydroxide is used up by 5 g of stearic anhydride.

Calculate the equivalent weight or the percentage of OH as usual.

### (h) *Acylation with 3,5-dinitrobenzoyl chloride* (*Robinson, Jr., et al.*[44]).

*Procedure:* For each series of analyses prepare fresh 0·2 M 3,5-dinitrobenzoyl chloride by dissolving 1·15 g with gentle heating in 25 ml pyridine. Do not expose the solution unnecessarily to moist air.

To determine liquid samples pipet approximately 4 m-equiv hydroxy compound into a tared 10-ml volumetric flask containing 3 ml pyridine. Reweigh and dilute to volume with pyridine.

Pipet into a 125-ml glass-stoppered Erlenmeyer flask 4·0 ml dinitrobenzoyl chloride solution, followed by 1·0 ml of the sample solution. Stopper tightly, swirl, and allow to stand for 5–15 min at room temp, then add 7–10 drops of water.

To determine solid samples, weigh accurately approximately 0·4 m-equiv directly into the 125-ml flask, then proceed as above.

Prepare a blank solution by pipetting 4·0 ml of the reagent solution into a flask, immediately adding 7–10 drops of water.

*Visual titration:* Add 40 ml pyridine to the reaction mixture, heat nearly to boiling, cool and then titrate with 0·2 N tetrabutylammonium hydroxide to the first definite and permanent red colour. Titrate the blank in exactly the same way. Calculate the equivalent weight of the compound from the difference in titre.

*Potentiometric titration:* Add 25 ml pyridine to the reaction mixture and heat nearly to boiling. Cool and transfer to a 250-ml electrolytic beaker. Rinse the flask with two 10-ml portions pyridine and add the washings to the beaker. Titrate potentiometrically under nitrogen. If a blank is determined, titrate only through the first inflexion, using the difference in volume between the end-points of blank and sample to calculate the equivalent weight (or per cent hydroxyl). If no blank is determined, titrate through both inflections and use the volume between the first and second end-points to calculate the equivalent weight. In the potentiometric titration proceed slowly through the first end-point to avoid a false inflexion.

The reactions are:

$$ROH + (NO_2)_2C_6H_3COCl + C_5H_5N \rightarrow (NO_2)_2C_6H_3COOR + C_5H_5N,HCl$$

$$HOH + (NO_2)_2C_6H_3COCl + C_5H_5N \rightarrow (NO_2)_2C_6H_3COOH + C_5H_5N,HCl$$

$$\text{First inflexion: } (NO_2)_2C_6H_3COOH + C_5H_5N,HCl$$

$$\text{Second inflexion: } (NO_2)_2C_6H_3COOR$$

When the 3,5-dinitrobenzoates have been isolated as derivatives for the characterization of the hydroxy compounds, their equivalent weight may be determined by titration with tetrabutylammonium hydroxide.

Dissolve 2–20 mg or more of the derivative in pyridine (Robinson, Jr. *et al.*[45]) or acetone (Smith and Haglund[46]), reflux for 30 min[45] and transfer the solution to an electrolytic beaker and titrate potentiometrically (glass electrode/calomel electrode) with 0·01 N[45] or 0·1 N[46] tetrabutylammonium hydroxide in benzene/methanol. Run a blank in parallel. From the difference between experiment and blank the equivalent weight of the ester is calculated as usual and then $E_{\text{Alcohol}} = E_{\text{Ester}} - 194{\cdot}2$.

### (*i*) *Acylation with pyromellitic dianhydride* (*Siggia et al.*,[47] *Kingston et al.*[48]).

*Reagent:* 0·1 M pyromellitic dianhydride in tetrahydrofuran.[47]

*Procedure:* Weigh accurately 1–1·5 m-equiv of the alcohol into a 250-ml flask. Add 25 ml reagent and 10 ml pyridine. Heat the flask on the steam-bath for 3 min and then on an electric hot plate for 8 min, most of the tetrahydrofuran boiling off during the heating. Add 20 ml pyridine to compensate, and continue the heating for further 5 min. Add 20 ml water and heat again for 3 min to hydrolyze excess anhydride. Cool to room temp and titrate with 0·1 N sodium hydroxide, indicator phenolphthalein. Run a blank in parallel.

From the reaction scheme:

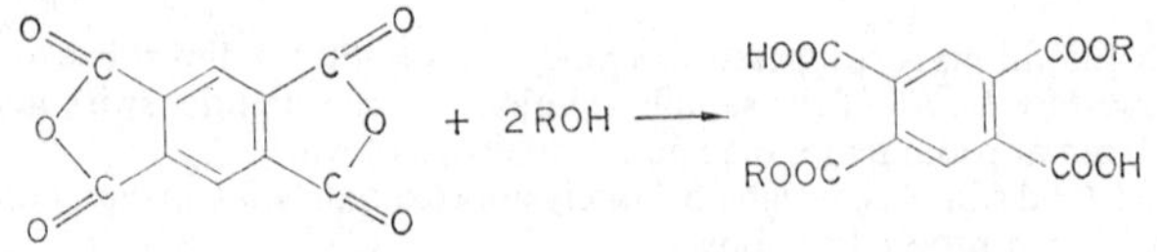


it is seen that 1 equiv of alcohol will cause a difference of 1 equiv of consumed base between blank and experiment.

*Reagent:* 0·19 M pyromellitic dianhydride in dimethylformamide.[48] Heat pyromellitic dianhydride at 160°C for 24 h and dissolve 41·0 g in 900 ml dimethylformamide. This solution can be stored for at least 1 month. Add before use 21 g imidazole and enough dimethylformamide to bring the total volume to 1000 ml.

*Procedure:* Weigh accurately 7–8 m-equiv of the hydroxy compound into a glass-stoppered 500-ml Erlenmeyer flask and pipet 50 ml reagent solution into the flask. Stopper the flask, swirl it and place it in a water-bath at 70–80°C for 5 min. Add 150 ml water and 5 drops of 2% solution of thymol blue in dimethylformamide. Titrate with 0·5 N potassium hydroxide until thymol blue gives no further colour change. Run a blank in parallel.

The reaction time can be shortened to 5 min at room temp by using a reagent solution containing 60 g pyromellitic dianhydride and 54 g imidazole in 1000 ml and weighing only 4–5 m-equiv of the hydroxy compound.

### (j) *Acylation with nitric acid in acetic anhydride (Schenk and Santiago[49]).*

*Reagents:* 0·42 M nitric acid. Pipet 11 ml acetic anhydride into 10 ml acetonitrile in a 100-ml volumetric flask. Cool the mixture in an ice-bath and keep it there while adding slowly, dropwise, 2·65 ml concentrated nitric acid. Then add about 50 ml acetonitrile and agitate the flask in the ice-bath for about 5 min. Allow the reagent to warm to room temp and dilute to the mark with acetonitrile. Keep the reagent away from direct sunlight or in the dark. The reagent should be prepared fresh daily.

0·6 N sodium hydroxide. To 240 ml saturated aqueous sodium hydroxide add 100 ml of water (free from carbon dioxide) and 5400 ml of absolute methanol (free from carbon dioxide). Standardize the titrant against potassium hydrogen phthalate.

*Mixed indicator:* Mix 3 parts of 0·1% neutralized thymol blue with one part of 0·1% neutralized cresol red.

*Procedure:* Weigh accurately a sample containing 2 m-equiv of primary hydroxy groups into a 125-ml glass-stoppered flask. Add by pipet exactly 10 ml reagent and swirl the flask gently. Allow the flask to stand for 20 min at room temp in the darkness or away from direct sunlight. Then add 2·5 ml water, swirl, and add 7·5 ml pyridine. Allow the flask to stand for 10 min at room temp, then add mixed indicator and titrate with 0·6 N sodium hydroxide to a colour change from yellow to violet.

Run a blank by pipetting exactly 10 ml nitric acid reagent into a glass-stoppered flask, and allow it to stand for 20 min at room temp in the same light as the sample. Then add 2·5 ml water and 7·5 ml pyridine and titrate with 0·6 N sodium hydroxide.

## 3. Arylurethane Formation

### *Method F. Reaction with Phenylisocyanate (Dreher,[50] Reed et al.[51])*

*Reagents[50]:*

0·5 N phenylisocyanate in chlorobenzene.

0·5 N diisobutylamine in chlorobenzene.

*Procedure:* Dissolve 1 g polyoxyalkylene compound in 10 ml chlorobenzene and acetone (1/1) and add excess phenylisocyanate reagent. The reaction is complete in 5 min at 20°C. Back-titrate the excess phenylisocyanate with diisobutylamine, using bromothymol blue indicator.

The reactions are:

$$HO(CH_2CH_2OCH_2CH_2)_nOH + 2C_6H_5NCO \rightarrow C_6H_5NHCOO(CH_2CH_2OCH_2CH_2)_nOOCNHC_6H_5$$

$$C_6H_5NCO + HN(C_4H_9)_2 \rightarrow C_6H_5NHCON(C_4H_9)_2$$

The equivalent weight is calculated as usual from the weight of polyoxyalkylene compound taken and the amount of phenylisocyanate used up (difference between a blank value and the amount of diisobutylamine used up in the experiment proper):

$$E = \frac{a \cdot 1000}{b \cdot n}$$

The mean molecular weight of the polyoxyethylene compound is then $2 \cdot E$, and the percentage of hydroxyl is $\frac{17 \cdot 100}{E}$.

The method recommended by Reed *et al.*[51] is more complicated, but it is recommended for serial estimations. Stannous octanoate is used as a catalyst for the urethane formation.

Otey *et al.*[52] preferred toluenediisocyanate for phenylisocyanate and triethylenediamine as catalyst. Excess isocyanate is determined by the addition of excess standardized diisobutylamine and back-titration of the excess with N hydrochloric acid.

## 4. Bromometric Estimation of Phenols

*Method G*

(*a*) (*Veibel*[53])

*Reagent:* An aqueous solution containing 2·8 g potassium bromate and 12 g potassium bromide per litre (i.e. 0·1 N in bromine), standardized against 0·1 N sodium thiosulphate.

*Procedure:* Weigh accurately about 0·2 g of the phenol into a 50-ml measuring flask and dissolve in water with the addition of sufficient sodium hydroxide to form the phenate. Pipet 10 ml of this solution into an Erlenmeyer flask equipped with a stopper carrying a tap funnel and a security funnel, the bulbs of which contain 10% potassium iodide acidified with a few drops of dilute sulphuric acid. Add 100 ml water and 25 ml reagent solution, insert the stopper and add through the tap funnel 20 ml 4 N sulphuric acid. Swirl the flask to mix the contents and leave it for 15 min at room temp. Excess bromine should be present, and if the solution remains colourless after acidification (the colourless bromination product usually precipitates) the experiment must be repeated, taking more than 25 ml reagent solution. Then add 20 ml 10% potassium iodide through the tap funnel, swirl the flask to allow any residual bromine vapour to react with the acid potassium iodide solution, rinse the tap funnel and the security funnel with water, adding the washings to the contents of the flask, and titrate the liberated iodine with sodium thiosulphate, indicator starch solution.

The most accurate results are generally obtained by using an excess of reagent solution corresponding to about 5 ml 0·1 N sodium thiosulphate. If too great an excess of bromine is used (an excess of more than 10 ml 0·1 N solution) products of the "tribromophenol bromide" type may be formed, causing a slow liberation of iodine with reappearance of the blue colour. Usually the iodine is titrated 5 min after the addition of potassium iodide, but if a rapid reappearance of the blue colour occurs it is advisable to wait 10 min before the iodine is finally titrated.

The equivalent weight per bromine atom consumed, calculated as usual, is multiplied by 2, 4 or 6, according to the number of *o*- and *p*-oriented hydrogen atoms present, to give the equivalent weight of the phenol.

### (*b*) *Smith*(54)

*Reagent:* As in *G* (*a*).

*Procedure:* Dissolve 1·5 m-equiv of phenol in 10 ml glacial acetic acid, add 5 ml concentrated hydrochloric acid and titrate with the reagent solution until the solution has assumed a distinct yellow colour. Add 10 ml 10% potassium iodide and titrate the liberated iodine with 0·1 N sodium thiosulphate.

When the reproducibility is poor the following procedure should be tried: weigh 1·5 m-equiv of the phenol into a glass-stoppered Erlenmeyer flask and add 10 ml glacial acetic acid, followed by 30 ml reagent solution and 5 ml concentrated hydrochloric acid. Stopper the flask immediately and leave it to stand for 3 min. Add 10 ml 10% potassium iodide solution and titrate the liberated iodine.

The equivalent weight of the phenol is calculated as in method *G* (*a*).

### (*c*) *Potentiometric micromethod for bromometric estimation* (*Chikryzova and Khomenko*(55)).

*Reagent:* As in Method *G* (*a*).

*Procedure:* Use a cathodically polarized platinum electrode at a polarizing current of $4{\cdot}6\times10^{-6}$ amp cm$^{-2}$. Weigh the sample accurately in a measuring flask, dissolve it in water and dilute to volume. To an aliquot containing $5\times10^{-6}$ to $1\times10^{-4}$ mole of phenol add 5–10 ml 50% sulphuric acid and dilute to 50 ml with water. Titrate potentiometrically with the reagent solution, stirring the solution magnetically. The potential jump is 500–800 mV.

### (*d*) *Electrometric estimation* (*Schoenberger*(56)).

Schoenberger found poor reproducibility in the amperometric and coulometric methods for bromometric estimation of phenols. He suggested the following procedure:

*Electrolytic cell:* Equip the cell with two platinum generating electrodes (the negative electrode is isolated by means of a tube ending in a capillary), 0·5 cm$^{-2}$ surface area; a polished platinum electrode and a S.C.E. are placed in the supporting electrolyte solution

(5 ml M sodium bromide and 5 ml M sulphuric acid in 200 ml of water). Operate at 2–3 m amp electrolytic current until the potential drop is 600 mV. Prepare the coulometer, add an exactly measured volume of the phenol containing 1–10 mg phenol litre$^{-1}$. Start the electrolysis anew and continue to a potential drop of 600 mV.

### (e) *Iodometric method, using iodine chloride* (*Karpov*[57]).

This method has been developed for the estimation of alkylphenols with branched side-chains.

*Reagent:* 0·1 N iodine chloride.

*Procedure:* Dissolve 0·3–1·2 g of the phenol in 50% ethanol (for heavily hindered phenols use 96% ethanol) and dilute to 50 ml with ethanol. To a 5-ml aliquot add 50 ml ethanol and 40 ml water, then 25 ml reagent solution and heat for 20 min (for heavily hindered phenols for 4 h). Then add 5 ml 20% potassium iodide and 100 ml water and titrate with 0·1 N sodium thiosulphate, indicator 3% starch in 50% glycerol. Compare with a blank. The error is claimed to be 0·15–0·20%.

### (f) *Bromination followed potentiometrically* (*Huber and Gilbert*[58]).

*Reagent:* 0·15 M solution of bromine in glacial acetic acid.

*Electrolytic cell:* Place two identical platinum foil electrodes (10 × 0·5 cm) 1·0 cm apart in the cell. Provide a constant current of 10 m amp by a 100 V d.c. source in series with a 10 MΩ resistor. Read the voltage across the electrodes with a vacuum tube voltmeter.

*Procedure:* Dissolve a sample of the phenol corresponding to about 20 ml reagent solution in 37 ml glacial acetic acid and 3 ml pyridine. Titrate the solution to a permanent voltage drop. Carry out a blank on the glacial acetic acid–pyridine mixture. The blank should not be more than 0·1–0·2 ml.

### (g) *Oxidation of alcohols with bromine chloride* (*Konishi et al.*[59]).

*Reagent and apparatus:* Dissolve 79·34 g potassium bromide and 55·67 g potassium bromate in water and dilute to 1000 ml. Addition of hydrochloric acid to this solution produces bromine chloride according to the scheme:

$$2KBr + KBrO_3 + 6HCl \rightarrow 3BrCl + 3KCl + 3H_2O$$

showing that the bromide–bromate solution behaves as a 1·0 M solution of bromine chloride.

The oxidations are carried out in a thermostat at 50 ± 1°C for primary alcohols, at 20 ± 1°C for secondary alcohols.

*Procedure:* Primary alcohols: By pipet, place 3 ml aqueous sample solution and 1 ml bromine chloride solution into a 100-ml iodine flask. Carry out the following steps in a dark room. Add 0·2 ml concentrated hydrochloric acid. Stopper tightly and seal with 4 ml 20% potassium iodide solution. Warm the mixture at 50°C in a thermostat for 3 h, then cool and allow the potassium iodide solution, sealing the stopper, to drop into the flask. Titrate the liberated iodine with 0·1 N sodium thiosulphate. Carry out a blank with

3 ml water. The amount of bromine chloride consumed in the oxidation is equal to the difference between iodine found in the blank and in the experiment.

The scheme

$$RCH_2OH + 2BrCl + H_2O \rightarrow RCOOH + 2HBr + 2HCl$$

shows that 1 mole of a primary alcohol uses up 2 moles of bromine chloride.

*Secondary alcohols:* The procedure, including the use of a dark room, is the same as for primary alcohols except for the oxidation temperature (20°C) and time (1·5 h).

The scheme

$$R_2CHOH + BrCl \rightarrow R_2CO + HBr + HCl$$

shows that 1 mole of a secondary alcohol uses up 1 mole of bromine chloride.

Concentration range 0·001–0·033 M for primary alcohols, 0·0025–0·05 M for secondary alcohols. Methanol cannot be determined by this procedure.

## 5. Oxidation Methods

### *Method H*

#### (*a*) *Oxidation with cerium(IV) hydrogen sulphate* (*Guilbault and McCurdy, Jr.*[64]).

*Reagents:* Treat 7·5 and 14·0 g of cerium(IV) hydrogen sulphate with, respectively, 5·0 and 7·0 ml of 98% sulphuric acid. Carefully heat the pastes for 5 min with stirring and dilute them gradually to 1000 ml with water containing 82·5 ml 72% perchloric acid. Standardize against standard arsenic(III) oxide.

*Standard iron(II) solution:* Dissolve 3·92 or 9·80 g of iron(II) diammonium sulphate, $6H_2O$ (Mohr's salt) in 1000 ml 0·5 N sulphuric acid. Standardize the solution every week against the cerium(IV) reagent, using ferroin as indicator.

*Catalyst A:* Dissolve 12·8 ml 50% solution of manganese(II) nitrate and 3·0 g silver nitrate in N perchloric acid to a total volume of 250 ml.

*Catalyst B:* Dissolve 12·5 g silver nitrate and 12·8 ml 50% solution of manganese(II) nitrate in N perchloric acid to a total volume of 250 ml.

*Procedure:* Dissolve 3–25 mg of the polyol in the cerium(IV) reagent (take 20–65% excess of the reagent), add 1 ml catalyst A solution and 8·5 ml 72% perchloric acid per 100 ml solution. Heat on a hot plate at 95°C until the solution turns red, then quench the reaction immediately by cooling and by the addition of 10–15 ml 6 N sulphuric acid. Titrate the excess oxidant with standard iron(II) solution, indicator ferroin.

It is indicated that all carbon atoms to which a hydroxyl group is attached are oxidized to formic acid according to the schemes:

$$\begin{matrix} CH_2OH \\ | \\ CH_2OH \end{matrix} + 6Ce(HSO_4)_4 + 2H_2O \rightarrow 2HCOOH + 3Ce_2(SO_4)_3 + 15H_2SO_4$$

$$\begin{matrix} CH_2OH \\ | \\ CHOH \\ | \\ CH_2OH \end{matrix} + 8Ce(HSO_4)_4 + 3H_2O \rightarrow 3HCOOH + 4Ce_2(SO_4)_3 + 20H_2SO_4$$

a diol thus requiring 6 and a triol 8 oxidation equivalents.

### (*b*) *Oxidation with gold(III) chloride* (*Srivastava and Saxena*[66]).

*Reagents:* 0·016 N gold(III) chloride, 0·8 N sodium hydroxide, 0·04 N potassium hexacyanoferrate(II), 0·01 N cerium(IV) sulphate.

*Procedure:* Weigh accurately 0·1–3 mg of the glycol and dissolve it in water. Add a known excess of the standard gold(III) chloride solution followed by an excess of standard sodium hydroxide. Keep the mixture on a hot plate for 3 h (reflux). Cool to room temp, remove the metallic gold by filtration, wash thoroughly with water, collect filtrate and washings, acidify, add a known excess of potassium hexacyanoferrate(II) and back-titrate with standard cerium(IV) sulphate, indicator *N*-phenylanthranilic acid, until a brownish-red colour appears.

From the scheme:

$$4\begin{array}{c}CH_2OH\\|\\CH_2OH\end{array}+4AuCl_3+12NaOH \rightarrow 4Au+8HCHO+12NaCl+10H_2O+O_2$$

it is seen that 1 molecule of glycol uses up 1 molecule (2 equiv) of gold(III) chloride, and that as a side-reaction, without analytical implication, gold(I) hydroxide will be converted to gold and free oxygen.

### (*c*) *Oxidation with xenon trioxide* (*Jaselskis and Warriner*[67]).

*Reagent:* Prepare an approximately 0·01 M xenon trioxide solution as described by Williamson and Koch[68] by hydrolysis of xenon hexafluoride or from xenon tetrafluoride according to the scheme

$$4XeF_4+8H_2O \rightarrow 2XeO_3+O_2+16HF+2Xe$$

which, however, is not followed stoichiometrically. The reagent is used directly without elimination of hydrofluoric acid. It is standardized iodometrically after the addition of sodium iodide by titration with 0·03 N sodium thiosulphate.

*Procedure:* To estimate the alcohol content of an aqueous solution pipet an aliquot containing not less than $5\times10^{-4}$ μmoles of alcohol into a 50-ml Erlenmeyer flask. Dilute to about 20 ml and add a known excess of xenon trioxide solution. Allow the mixture to stand for 2 h, then add several drops of 2 N sulphuric acid and excess sodium iodide. Titrate the liberated iodine with standard sodium thiosulphate.

The amount of alcohol present is calculated either from a calibration curve characteristic of each alcohol or by the solution of an equation equally characteristic of each alcohol.

The calibration curves are rectilinear but with an intercept on the ordinate axis ($XeO_3$), the mean value of which is $1{\cdot}125\times10^{-4}$ m-mole. The slope of the line is characteristic for the alcohol.

The equations are of the type $y = a+m\cdot x$, $y$ being m-moles of xenon trioxide, $a$ the intercept with the $y$-axis, $m$ the slope and $x$ m-moles of alcohol in the sample.

For methanol, ethanol, and propan-2-ol, $y$ = 1·116, 1·145, 1·115 and $m$ = 1·204, 2·091, 2·977, respectively.

*(d) Oxidation with diphenylpicrylhydrazyl (Papariello and Janish*[70]*).*

*Reagent:* $10^{-4}$ M solutions of diphenylpicrylhydrazyl and of diphenylpicrylhydrazine in absolute ethanol are prepared by stirring solvent and solute for 30–35 min and after filtration diluting to the desired concentration.

Add 25 ml reagent solution to 2 ml $2{\cdot}4 \times 10^{-4}$ M solution of the phenol and dilute to 50 ml with ethanol. Take absorbance readings at 540 nm at appropriate times, using a $5 \times 10^{-5}$ M solution of diphenylpicrylhydrazine to set 100% transmission. Keep all reaction mixtures in the dark between readings to avoid the slow photochemical reduction of the hydrazyl radical to hydrazine which occurs in sunlight at 520 nm.

The decrease in absorbance is proportional to the amount of monohydric or dihydric phenol present, each mole of monohydric or dihydric phenol reducing, respectively, 1 or 2 moles of hydrazyl radical to hydrazine.

A calibration curve should be established for each phenol considered.

Water affects the rate of the reaction and should therefore be avoided.

## 6. Elimination of Water and Karl Fischer Titration

### *Method I (Bryant et al.*[72]*)*

*Reagent (catalyst solution):* Dissolve about 100 g boron trifluoride gas and 1 or 2 ml water in sufficient glacial acetic acid to make 1 l. solution.

*Procedure:* Weigh 5–10 mg sample accurately into a 100-ml volumetric flask, about one-third filled with dioxan; make up nearly to volume with this solvent, shake until homogeneous, and adjust to volume after placing the flask for at least 30 min in a water bath at 25±1°C.

Transfer 5 ml solution to a 250-ml volumetric flask and add 20 ml catalyst solution. Stopper the flask tightly with a spring clamp and place it together with a blank (5 ml dioxan and 20 ml catalyst solution) in an oven or a water-bath at 67±2°C for 2 h. Allow to cool to room temp spontaneously, then add 5 ml pyridine and titrate with Karl Fischer reagent.

After correction for water in the dioxan, catalyst and pyridine, the m-moles of water present, less the water originally present, are equal to the m-equiv of free hydroxyl groups in the sample.

For the determination of water originally present titrate 5 ml of the dioxan solution with Karl Fischer reagent.

## 7. Estimation of Active Hydrogen

### *Method J (Norton et al.*[79]*)*

The reaction vessels are shown in Fig. 9.

*Apparatus:* Gas chromatograph, suitable for operation under the following conditions: columns 900 mm × 4 mm i.d. copper tube packed with molecular sieves 5A, 60–80 B.S. mesh. Temperature: detector 100°C, column ambient; recorder, 1 mV, 1 sec, with chart speed 120 in. hour$^{-1}$; detector, thermal conductivity; and carrier-gas, nitrogen, with cylinder inlet pressure 40 lb in$^{-2}$, outlet pressure atmospheric, and flow-rate 60 ml min$^{-1}$ (adjusted before reaction cell).

Reaction vessel. This was a glass cell, fitted with a side-arm for the inlet of carrier-gas and connected to the chromatograph by B7 Quickfit joint (Fig. 9).

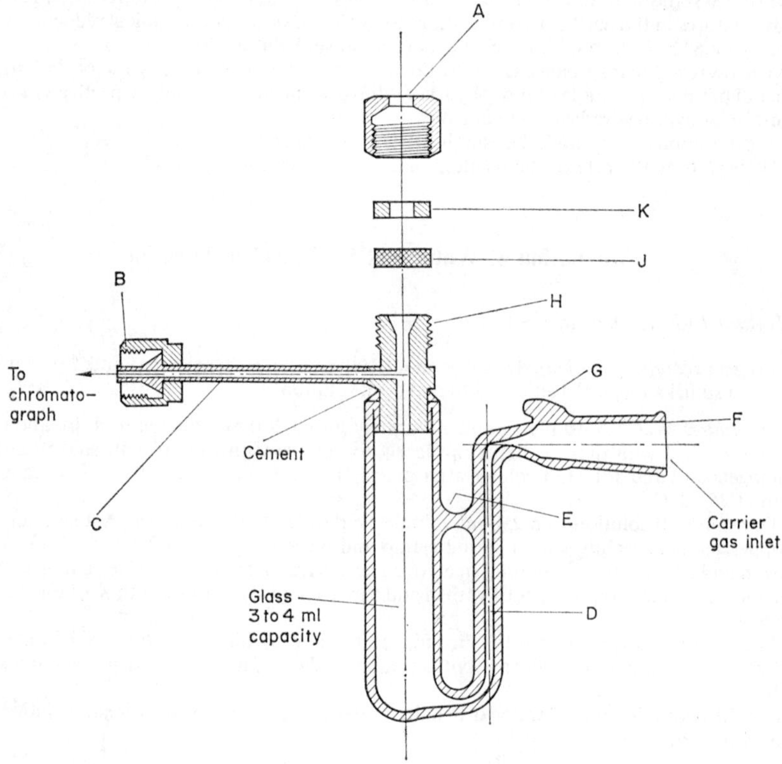


FIG. 9. Gas chromatographic estimation of active hydrogen (Norton, Turner and Salmon[79]).

A. Injection port
B. Stainless steel Swagelock coupling
C. $\frac{1}{8}$ in. o.d. stainless steel side arm brazed on
D. Capillary
E. Brace
F. B7 joint
G. Glass hook for retaining spring
H. $\frac{1}{8}$ in. Wade stainless steel double-ended union drilled to $\frac{1}{8}$ in. bore
J. Septum
K. Spacer washer

(Reprinted by permission of the Society for Analytical Chemistry and the authors.)

Sampling valve. This was a four-port, two-way valve, which allows carrier-gas to pass to the chromatograph either via the reaction vessel or directly (Fig. 10).

Syringe. These were 100 μl Hamilton, 2 ml and 10 ml with wide-bore needle.

Sample vials. Glass vials of about 15 ml capacity, fitted with plastic caps. Rubber septums were fitted over the mouths of the vials.

Reflux system. A 150-ml conical flask, fitted with a water-cooled condenser and closed with a drying tube, and a hot plate.

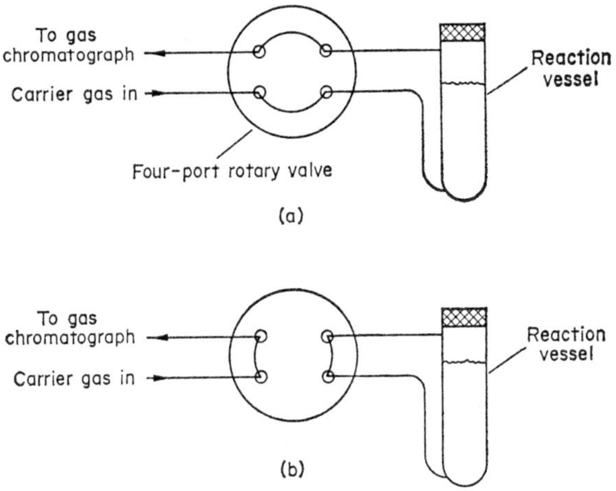


FIG. 10. Details of flow system for the apparatus shown in Fig. 9. (a) Sampling position, (b) Isolated position. (Reprinted by permission of the Society of Analytical Chemistry and the authors.)

*Reagents:* Tetrahydrofuran. Technical grade, dried over molecular sieves for at least 1 week.

Lithium aluminium hydride solution. Reflux about 2 g lithium aluminium hydride (laboratory grade) with 100 ml dried tetrahydrofuran in the reflux apparatus for about 1 h, maintaining gentle reflux. Disconnect the flask and fit it with the drying-tube. Cool to obtain a clear solution with a grey sediment consisting of undissolved reagent. Transfer the clear solution to a clean, dry 50-ml flask with a 10-ml syringe, and stopper the flask immediately.

Molecular sieves. Grade 5A, 60–80 B.S. mesh, activated by heating at 250°C for 2 h.

Benzoic acid. Analytical-reagent grade.

*Procedure:* Weigh accurately into sample vials amounts of sample and of benzoic acid. Stopper the vials with the plastic caps.

Transfer by pipet 5 ml dried tetrahydrofuran into each vial, stopper and shake each vial until the sample is dissolved. Add 10–15 particles of activated molecular sieves to each vial and allow them to stand in a desiccator for 30 min to remove traces of water.

Connect the clean dry reaction vessel to the gas chromatograph sampling manifold. Introduce 2 ml lithium aluminium hydride reagent into the reaction vessel with a syringe. Screw on the septum and injection port.

Turn the sampling valve so that the carrier-gas bubbles through the reagent, thus flushing any hydrogen into the chromatograph. After about 10 sec (before the hydrogen peak has emerged) turn the valve so that the carrier-gas passes directly to the chromatograph. Repeat the valve movements in this way until no hydrogen peak is obtained. Should a small surge peak of constant height be obtained this can be ignored, as a correction for it will be made in the blank run.

Remove the plastic cap from the vial containing the sample solution and rapidly cap the vial with a rubber septum. Fill a 100-μl syringe with sample solution *via* the rubber septum. Inject 50 μl of sample solution into the reaction vessel with the syringe through the rubber septum situated in the inlet port, taking care that the contents of the syringe pass directly into the hydride reagent solution without touching the sides of the vessel. An instantaneous evolution of hydrogen bubbles takes place and after about 1 min the reaction is complete. Switch the sampling valve to the sampling position to sweep the hydrogen formed into the chromatograph. Measure the area of the hydrogen peak either by electrical integration or by cutting out the peak and weighing.

Carry out a duplicate determination by repeating the cell flushing procedure and introduce the sample solution as previously described. Three or four replicate determinations can be carried out with the same charge of hydride reagent. After this the reagent becomes gelatinous and it is not possible to effect quantitative hydrogen recovery. Carry out a determination in a similar manner with the benzoic acid solution as standard.

After about thirty injections the column should be cleansed from any tetrahydrofuran that slowly passes to it during the analysis. This can be done by raising the column temp to 150°C and passing carrier-gas through the column for 1 h, or preferably overnight.

Carry out a blank determination on 50 μl of the dried tetrahydrofuran in the absence of either sample or benzoic acid. Clean the reaction cell before carrying out further determinations.

To clean the reaction vessel, disconnect it from the manifold and remove the injection port. Wash the vessel with water followed by dilute hydrochloric acid. Rinse with distilled water and acetone and dry in an oven at 105°C before reassembling.

*Calculation:* Calculate the volume of hydrogen, $R$, represented by unit area of the hydrogen peak from the data obtained when using benzoic acid as follows:

$$R = \frac{11200\,wv}{122t(a-b)} \text{ ml of hydrogen per unit area,}$$

where $w$ is the weight (g) of benzoic acid taken, $a$ the number of area units of the benzoic acid peak, $b$ the number of area units of the blank peak, $v$ the volume (μl) of benzoic acid solution injected and $t$ the total volume of benzoic acid solution, i.e. 5000 μl.

Calculate the hydroxyl content of the compound as follows:

$$\text{Hydroxyl content (as OH equivalents per 100 g)} = \frac{100Rt_1(a_1-b)}{11200w_1v_1}$$

where $R$ is the volume (ml) of hydrogen per unit area (as calculated above), $w_1$ is the weight (g) of compound taken, $a_1$ is the number of area units of the compound peak, $b$ is the number of area units of the blank peak, $v_1$ is the volume (μl) of sample aliquot, and $t_1$ is the total volume of sample solution, i.e. 5000 μl.

## 8. Titanometric Estimation of Nitro Group-containing Derivatives

*Method K. Titration with Titanium(III) Chloride (Kolthoff and Robinson*[86]*)*

*Reagent:* 0·05–0·1 N titanium(III) chloride in 2–3 N hydrochloric acid. Mix 200 ml commercial 15% solution of titanium(III) chloride in hydrochloric acid with 400 ml concentrated hydrochloric acid and dilute the mixture with water to a total volume of 2 l. Keep this oxygen-free by passing a stream of oxygen-free nitrogen or carbon dioxide through the solution, very slowly when the solution is not in use, more rapidly when it is used. This stream is freed from oxygen by passing it through two wash-bottles each containing 5% titanium(III) chloride in 30% aqueous sodium citrate. The water-vapour content of the gas stream is adjusted by passing it through a wash-bottle with water. The assembly is shown in Fig. 11.

The solution is standardized against 0·06 N iron(III) ammonium sulphate, using ammonium thiocyanate as indicator.

*Procedure:* As each nitro group requires 6 equiv of titanium(III) for reduction to the amine stage only 10–20 mg of the nitro compound should be used for an estimation. Dissolve 0·2 g sample in ethanol in a 50-ml measuring flask. Place an aliquot (5 or 10 ml) of this solution in an Erlenmeyer flask (300 ml), add 0·5 g sodium hydrogen carbonate and 30 ml of 20% aqueous disodium hydrogen citrate, thus adjusting the reaction of the solution to about pH 7. The reduction potential of titanium$^{3+}$ is much higher at neutral reaction than in acid solution. Partial precipitation of the nitro compound may occur, but the precipitate is usually so finely divided that it does not interfere with the titration, passing into solution as the reaction proceeds.

The Erlenmeyer flask is equipped with a stopper fitted with two glass tubes, one reaching the bottom of the flask, and the other—wide enough to allow the tip of the burette to pass through it—ending just below the stopper. A stream of oxygen-free carbon dioxide or nitrogen is passed through the flask for 3 min to remove dissolved oxygen. Titanium(III) chloride is then added from a burette through the wide tube, the stream of carbon dioxide (or nitrogen) being continued, until the colour of the solution is blue-violet (in some instances brown), too great an excess being avoided. The titanium(III) chloride is strongly acid, and its addition will cause a vigorous evolution of carbon dioxide due to the presence of sodium hydrogen carbonate in the solution, which is thus most effectively freed from the last traces of oxygen.

The flask is left to stand for 2 min, the stream of carbon dioxide (nitrogen) being continued; then 20 ml concentrated hydrochloric acid are added through the wide tube. The stream of carbon dioxide (nitrogen) is discontinued, the glass tubes rinsed with a few ml water, 5 ml 10% ammonium thiocyanate are added and the excess titanium$^{3+}$ titrated with 0·06 N iron(III) ammonium sulphate until the red colour of iron(III) thiocyanate persists for at least 20 sec. If the ammonium thiocyanate is added before the concentrated hydrochloric acid it may be reduced by titanium(III) ions, thus invalidating the estimation.

A blank is carried out to eliminate any consumption of the reagent by the sodium citrate solution.

When $a$ g of substance have been dissolved to 50 ml and $x$ ml are taken for the estimation, the equivalent weight, $E$, of the compound, i.e. the weight of the compound using up 1 equivalent of titanium(III), is:

$$E = \frac{a \cdot 1000 \cdot x}{(b-c) \cdot n \cdot 50}$$

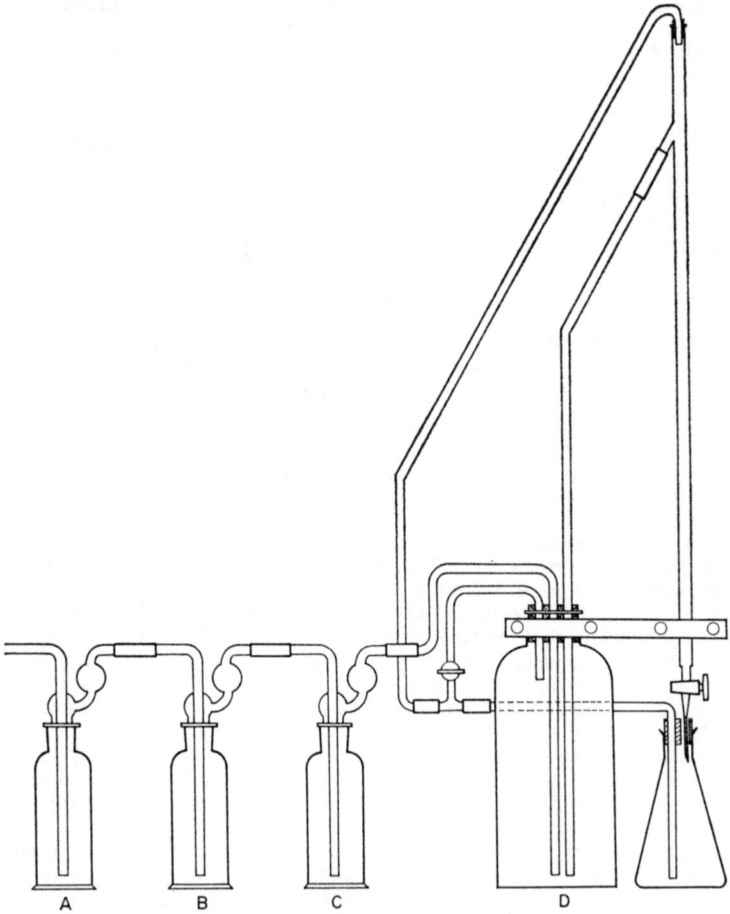


FIG. 11. Assembly for titanometric estimation of nitro groups. A and B. Wash-bottles containing 5% titanium(III) chloride in 20% disodium hydrogen citrate; C. Wash-bottle with water; D. 0·05–0·1 N titanium(III) chloride in hydrochloric acid. Kolthoff and Robinson.(86)

where $b$ is the volume of iron(III) ammonium sulphate of normality $n$ equivalent to the volume of titanium(III)-reagent added, and $c$ is the volume of iron(III) ammonium sulphate used up in the back-titration.

For a mono-nitro compound the molecular weight, $M$, is then 6 $E$; for a dinitro compound $M$ is 12 $E$.

Instead of the apparatus shown in Fig. 11 Siggia[182] recommends the use of an apparatus as used in the Karl Fischer titration of water. This arrangement is shown in Fig. 12 and Fig. 13.

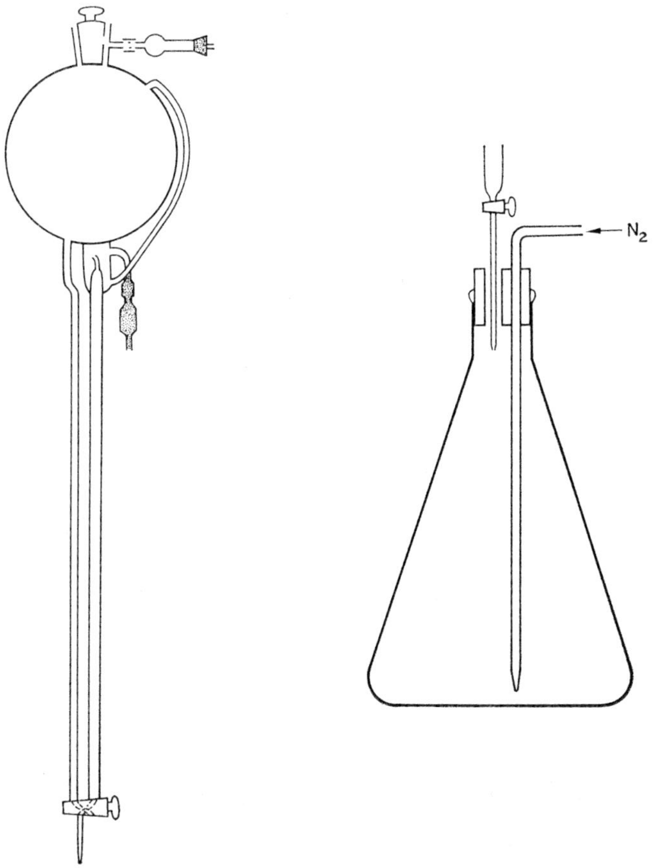


FIG. 12. Karl Fischer burette adapted to titanometric estimations (Siggia[182]). (Reprinted by permission of John Wiley & Sons Inc. and the author.)

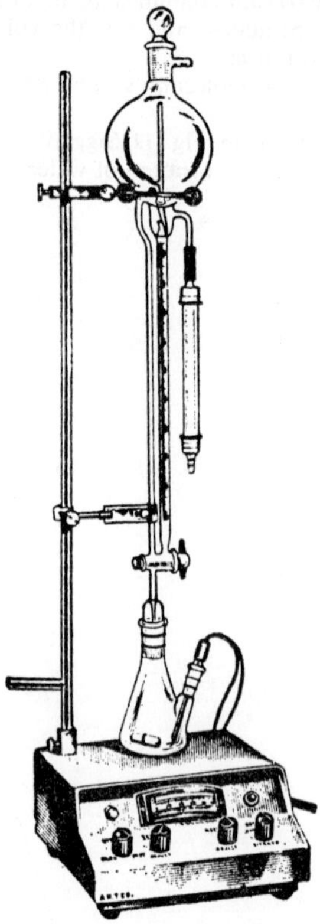

FIG. 13. Karl Fischer titration unit, adaptable to titanometric estimations. (Courtesy Arthur H. Thomas Co., Philadelphia.)

## 9. Complexometric Estimation of Hydroxy Compounds

*Method L. Vanadium Oxinate Method*

(*a*) *Pesez and Bartos*[91]

*Reagents:* Dissolve 0·27 g ammonium vanadate in 7 ml N sodium hydroxide; dilute to 10 ml with water and boil for 5 min to remove ammonia.

In a flask, dissolve 1 g 8-hydroxyquinoline in 3·5 ml 2 N sodium hydroxide (heating may be necessary). Dilute with water to 15 ml, heat on a boiling water-bath and add, with mechanical stirring, the sodium vanadate solution, continuing the heating and stirring for a further 10 min. Add in drops to the hot solution 50% acetic acid until the dark precipitate remains undissolved (about 1 ml). Filter the hot solution and cool the filtrate in an ice-bath. The sodium salt precipitates as yellow crystals. Isolate the salt by filtration, wash with water and dry *in vacuo* over potassium hydroxide. Yield 0·92 g.

Chloroform is freed from ethanol by washing 500 ml chloroform with 3 × 100 ml of 1 % potassium dichromate in N sulphuric acid, each time shaking the separatory funnel intermittently for 15 min, then wash with 100 ml water, 100 ml N sodium hydroxide, 4 × 100 ml water, and dry finally over magnesium sulphate.

*Procedure:* Suspend 0·04 g reagent in 9 ml purified chloroform, heat to boiling, and add 1 ml chloroform containing 0·026 g glacial acetic acid per ml. Boil for 1 min, cool to 20°C, and filter (practically no residue should be found). The solution is prepared immediately before use.

Dissolve the sample in 4 ml purified chloroform and add 1 ml reagent solution. Leave the mixture to stand for 2·5 h at 20–22°C, then add 5 ml N sodium hydroxide and shake until the black-violet colour of the reagent has disappeared. Decant and collect in a volumetric flask the chloroform phase (which should be limpid, red-orange in the experiment and colourless or very faint yellow in the blank). Add 0·5 ml glacial acetic acid containing 5 % dichloroacetic acid, adjust to the mark with chloroform and read the optical density at 600 nm. The colour is stable for 1 h at 20–22°C.

### (*b*) *Stiller*[92]

*Reagents:* A solution of ammonium vanadate containing 1 mg of vanadium per ml. 2·5 % solution of 8-hydroxyquinoline in 6 % acetic acid.

About 2 M acetate buffer (pH 5, 100 ml glacial acetic acid and 53·5 g sodium acetate dissolved in water to 1000 ml).

Benzene, reagent grade.

*Procedure:* In a separatory funnel add 1 ml ammonium vanadate solution to a sample of the substance or solution under test (containing 50–500 μg of alcohol), followed by buffer solution to a volume of 7 ml. Add 15 ml benzene (the two phases should be kept at fixed relative volumes, the ratio of the organic phase to the aqueous phase being approximately 2:1); shake the contents for 20 min, separate the phases and repeat the extraction from the aqueous phase twice with fresh quantities of benzene and the other reagent. Collect the benzene extracts, wash with 10 ml N sodium hydroxide, adjust with benzene to the mark in a volumetric flask and determine the optical density at 390 nm, comparing with a calibration curve for each alcohol.

## 10. Colorimetric Estimation of Hydroxy Compounds

### *Method M. Iron(III) Hydroxamates (Mizukami and Nagata,*[98] *Gutnikov and Schenk*[99]*)*

*Reagents:* Acetylating reagent: By pipet, add 1 ml 72 % perchloric acid to a 50-ml volumetric flask and immerse in ice water. By pipet, add slowly 20 ml pyridine down the sides of the flask, then dilute to the mark with acetic anhydride. Prepare the acetylating reagent fresh daily.

Hydrolysis reagent: Mix 4 volumes water with 1 volume pyridine.

1·4 M hydroxylammonium perchlorate: Dissolve 195 g sodium perchlorate in 450 ml absolute methanol with heating. Dissolve 105 g of reagent grade hydroxylammonium chloride in 550 ml absolute methanol with heating and stirring. Add the methanolic solution of sodium perchlorate slowly to the latter solution with stirring. Add 50 ml benzene and allow to cool to room temp, then chill the solution and the precipitate for 1 h in an ice-bath, filter the sodium chloride off on a sintered-glass suction filter and store the hydroxylammonium perchlorate solution in a polyethylene bottle.

2·5 M sodium hydroxide in absolute methanol.

Stock iron(III) perchlorate solution: Dissolve 50 g anhydrous iron(III) perchlorate in 400 ml absolute methanol, add 40 ml 72% perchloric acid and dilute to exactly 500 ml with ethanol.

Final iron(III) perchlorate reagent: To about 1·7 l. absolute ethanol add 100 ml stock iron(III) perchlorate solution. Add slowly and with cooling 140 ml of 72% perchloric acid in small portions. Allow to cool and dilute to 2 l. with absolute ethanol.

*Procedure:* Weigh accurately a hydroxyl-containing sample into a 10-ml (or larger) volumetric flask so that the solution contains 0·005–0·007 m-moles of hydroxy per ml. Dilute to the mark with the acetylating reagent, mix, and allow to stand for 5 min, or longer for hindered compounds. The ratio of sample to the total volume at this point should be no more than 1:20. Pipet exactly 1 ml of this solution into a 50-ml volumetric flask (for the blank use 1 ml of the acetylating reagent), add 0·5 ml of the hydrolysis reagent, mix, and allow to stand at room temp for 10 min. By pipet, add 3 ml 1·4 M hydroxylammonium perchlorate and then 10 ml of 2·5 M sodium hydroxide. After 20 min add about 22–23 ml of an iron(III) perchlorate–acetone mixture made by mixing, for each sample, 2·5 ml acetone with 35 ml of iron(III) perchlorate reagent. Allow to stand for 5 min, dilute to the mark with the iron(III) perchlorate–acetone mixture, mix thoroughly, and read the absorbance of the sample at 524 nm against a blank treated in the same manner.

The method has been used for the determination of hydroxyl in various polyglycols.

## 11. Complexometric Estimation of Hydroxy Compounds

### *Method N. Complex Formation with Ammonium Hexanitratocerate(IV) (Obtemperanskaya et al.*[100, 101]*)*

*Reagent:* 0·30–0·35 M diammonium hexanitratocerate(IV).

Dissolve 165–190 g diammonium hexanitratocerate(IV) in 4 N nitric acid to 1000 ml. Standardize the solution against 0·1 N Mohr's salt in the presence of ferroin.

*Procedure:* In a 25-ml volumetric flask, dilute a weighed amount of the hydroxy compound with a few ml water (if insoluble in water add just sufficient dioxan to keep the compound dissolved), add 2 ml of the reagent and fill up to the mark with water or water–dioxan mixture. Measure the absorbance in a colorimeter using a 5-cm cell and a blue filter. Compare with the absorbance of a blank and find the concentration of the hydroxy compound from a calibration curve.

## 12. Fluorimetric Estimation of 1,2-Diols

### *Method O* (*Pesez and Bartos*[102])

*Reagents: a.* 0·05 N sodium periodate;
*b.* 2·35 g tin(II) chloride, $2H_2O$ in 5 ml concentrated hydrochloric acid, diluted to 100 ml with water.
*c.* 4% (v/v) ethyl acetoacetate in 20% ammonium acetate.

*Procedure:* To 1 ml of a solution of the diol add 0·5 ml reagent *a* and, after 30 min at room temp, 0·5 ml reagent *b* followed by 2 ml water and 1 ml reagent *c*, stirring after each addition. Heat to 60°C for 20 min, cool, and filter. Determine in a spectrofluorimeter the extinction at 365 nm, and the emission at 470 nm.

Standardize against a solution of 3,5-diethoxycarbonyl-1,4-dihydrolutidine.

The reactions are:

$$RCHOHCHOHR' + HIO_4 \longrightarrow RCHO + R'CHO + HIO_3$$

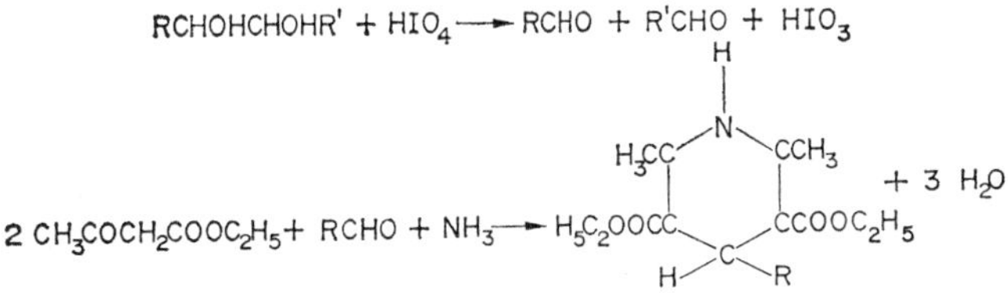

## 13. Estimation of Alcohols through Formation of 4-Amino-1,1′-azonaphthalene

### *Method P* (*Hoare et al.*[113])

*Reagents:* Acetate buffer. 250 ml glacial acetic acid and 125 g sodium acetate made up to 500 ml with water.

Sodium nitrite. 40 g sodium nitrite in 100 ml water.

α-Naphthylamine. 0·0625 g α-naphthylamine (recrystallized from boiling light petroleum) dissolved in 50 ml pure ethanol, containing 8·8 ml 10 N hydrochloric acid $l^{-1}$, made up freshly every 2 d and kept in darkness.

*Procedure:* Add 2 ml acetate buffer and 1 ml sodium nitrite to 8 ml of the alcoholic solution in a 50-ml volumetric flask. Stopper the flask tightly and leave it for 30 min in a water-bath at 18°C. Remove the stopper for a minimum of time and add 5 ml heptane; shake the flask mechanically at 18°C for 10 min and then cool to 0°C in an ice-bath. (The following procedures and apparatus were kept at 0°C as far as possible.) Transfer the solution to a previously cooled separatory funnel and run off the aqueous layer. Pipet the heptane to a precooled test-tube containing ten pieces of calcium chloride dried in an oven prior to use. Dry for 10 min at 0°C and repeat the operation with fresh calcium chloride.

Pour the heptane layer from the calcium chloride into a flask containing 4 ml water and 5 ml α-naphthylamine reagent. Cork the flask and allow it to stand at 18°C for 10 min; then place it in a water-bath at 60°C for 30 min for the colour to develop fully. Measure the absorbance against water at 550 nm in 1-cm glass cells.

The method is sufficiently sensitive to measure $10^{-7}$ moles of methanol $ml^{-1}$ of aqueous solution.

A calibration curve can be established for any alcohol.

## 14. Estimation of Tertiary Alcohols through U.V. Spectra of Alkyl Iodides

### *Method Q*

#### (*a*) (*Scoggins and Miller*[122])

*Reagents:* Commercial hydriodic acid 50–55%, or iodine-free 72%.

*Instrumentation:* Cary recording spectrophotometer, model MS 11, equipped with 1-cm quartz cell.

*Procedure:* Transfer a quantity of the sample (dissolved in a hydrocarbon) containing not more than 0·08 m-mole of tertiary alcohol to a 40-ml screw-cap vial equipped with a polyethylene gasket in the cap and dilute to 25 ml with cyclohexane. (For aqueous solutions, weigh 100 to 200 mg of sample into the vial and add 25 ml cyclohexane.) Add 3 ml hydriodic acid and shake for 3 min. Transfer the contents of the vial to a separatory funnel, rinse the vial with 10 ml water and add the rinse water to the separatory funnel. Mix thoroughly, allow the phases to separate, and discard the aqueous layer. Add 10 ml M sodium hydroxide and 3 drops 30% hydrogen peroxide and shake until the iodine colour disappears. Transfer an aliquot of the hydrocarbon phase to a 1-cm cell and measure the absorbance versus a reagent blank by scanning from 300 to 250 nm (ethyl iodide has maximum absorbance at 268 nm). Calibration curves (one for each tertiary alcohol) are prepared separately.

Tertiary alcohols up to $C_{10}$ have been measured from the ppm to 50% level in the presence of a variety of oxygenated materials, including primary and secondary alcohols. The method is applicable to both aqueous and hydrocarbon samples.

#### (*b*) *Bradley and Penketh*[60]

These authors use the technique just described for converting the tertiary alcohol into alkyl iodide, but prefer to eliminate hydrogen iodide from the alkyl iodide, convert it into iodine by reaction with iodic acid, extract the iodine into cyclohexane, and measure the absorbance at 525 nm.

*Procedure:* Follow the instructions of Scoggins and Miller[122] until the iodine has been destroyed by hydrogen peroxide, then separate the phases, discard the aqueous phase, wash the cyclohexane layer with 2 × 10-ml portions of distilled water, discard the washings, and filter the cyclohexane layer through a silicone-treated filter into a 150-ml Erlenmeyer flask. Rinse the separating funnel and the filter paper with 2 × 5-ml portions of cyclohexane, combine the rinsings with the contents of the flask, add to the flask 10 ml 30% potassium hydroxide and 10 ml ethanol and heat under reflux for 30 min. Allow to cool and wash the condenser with 20 ml water, running the washings into the flask. Separate the phases in a separatory funnel, washing the flask with sufficient water to form an aqueous phase of at least 50 ml to ensure a good phase separation. Discard the organic phase, acidify the aqueous phase with 10% sulphuric acid, cool and add 3 ml of

4% potassium iodate solution. Extract the liberated iodine into 25·0 ml cyclohexane and measure the absorbance at 525 nm, using a 1-cm cell.

Calibrate with a pure tertiary alcohol or with standard potassium iodide solution following the extraction procedure.

## 15. Gas Chromatographic Estimation

### *Method R*

*(a) Estimation of phenols through 2,4-dinitrophenyl ethers* (*Cohen et al.*[168])

*Instrumentation:* Silicon GE XE-60 columns (Cohen and Wheals[169]).

Preparation of derivatives from microgram quantities of phenol:

(i) Pipet 4 ml acetone containing the phenol, 0·1 ml of saturated sodium methoxide solution in methanol and 1 ml of 1-fluoro-2,4-dinitrobenzene solution (1% w/v in acetone) into a 10-ml pear-shaped flask and reflux for 30 min. Add the solution to 25 ml of aqueous 2·5% (w/v) sodium hydroxide, wash in with a little water and extract the mixture with 25 ml chloroform. Dry the chloroform layer with anhydrous sodium sulphate and evaporate to a low volume. Dissolve the residue in acetone, and inject a 5-μl aliquot into the gas chromatograph.

(ii) Phenols in aqueous solution. For etherification, follow the indications given in Part B, 6.1 (*b*), p. 37 and recommended Methods *L–O*, p. 61. Extract the ether with hexane, then proceed as indicated above.

(iii) Phenols separated by thin-layer chromatography or by paper chromatography. Etherify the phenols on the plate or paper by spraying with 4% acetone solution of 1-fluoro-2,4-dinitrobenzene (7 ml 400 $cm^{-2}$; sandwich the layer or the paper between two glass plates clipped tightly together, and heat in an oven at 190°C for 40 min. When cool, scrape off the interesting areas of the layer, transfer them separately to short columns plugged with cotton wool and eluate the ethers with 2 × 5-ml portions of acetone. From paper chromatograms, cut out the spots of interest, then treat them in a tube with 10 ml of acetone together with 2 drops water. Inject a 5-μl aliquot into the gas chromatograph.

*(b) Estimation of hydroxy compounds through gas chromatography of trimethylsilyl ethers*

The procedures described by Langer *et al.*,[170] Freedman and Croitoru,[171] Sweeley *et al.*,[172] Dutton *et al.*[174] and van Ling[175] (see pp. 108–109) are recommended as fast, reliable methods for the estimation of hydroxy compounds.

# REFERENCES TO PART C


1. van der Heijde, H. B. and Dahmen, E. A. M. F., *Anal. Chim. Acta* **16**, 378 (1957).
2. van der Heijde, H. B., *Anal. Chim. Acta* **16**, 392 (1957).
3. van der Heijde, H. B., *Anal. Chim. Acta* **17**, 512 (1957).
4. Streuli, C. A., *Anal. Chem.* **32**, 407 (1960)
5. Crabb, N. T. and Critchfield, F. E., *Talanta* **10**, 271 (1963).
6. Fritz, J. S. and Keen, R. T., *Anal. Chem.* **25**, 179 (1953).
7. Coetzee, J. F. and Padmanabhan, G. R., *J. Phys. Chem.* **69**, 3193 (1965).
8. Morman, D. H. and Harlow, G. A., *Anal. Chem.* **39**, 1869 (1967).
9. Harlow, G. A., *Anal. Chem.* **34**, 148 (1962).
10. Sensabaugh, A. J., Cundiff, R. H. and Markunas, P. C., *Anal. Chem.* **30**, 1445 (1958).
11. Harlow, G. A. and Wyld, G. E. A., *Anal. Chem.* **34**, 172 (1962).
12. Harlow, G. A. and Bruss, D. B., *Anal. Chem.* **30**, 1833 (1958).
13. Buell, B. E., *Anal. Chem.* **39**, 762 (1967).
14. Diembeck, S., Grauer, E. and Jaeger, H., *Arch. Pharm.* (*Weinheim*) **301**, 628 (1968); *Chem. Abs.* **69**, ref. 102880 (1968).
15. Fritz, J. S. and Gainer, F. E., *Talanta* **15**, 939 (1968).
16. Harlow, G. A., Noble, C. M. and Wyld, G. E. A., *Anal. Chem.* **28**, 787 (1956).
17. Belcher, R., Dryhurst, G. and Macdonald, A. M. G., *Anal. Chim. Acta* **38**, 435 (1967).
18. Aksenenko, V. M., Aksenenko, E. G. and Gromova, E. N., *Zavodsk. Lab.* **32**, 19 (1966); *Chem. Abs.* **64**, 18407 (1966).
18a. Fialko, M. M., Balashova, A. N. and Lyashenko, A. F., *Zh. Anal. Khim.* **25**, 1648 (1970); *Chem. Abs.* **74**, ref. 27859 (1971).
19. Greenhower, E. J. and Smith, J. W., *Analyst* **84**, 457 (1959).
20. Pandya, P. K. and Haldar, B. C., *J. Sci. Ind. Res. India* **21B**, 503 (1962); *Chem. Abs.* **58**, 2850 (1963).
21. Vaughan, G. A. and Swithenbank, J. J., *Analyst* **90**, 594 (1965).
22. Vaughan, G. A. and Swithenbank, J. J., *Analyst* **95**, 890 (1970).
23. Fritz, J. S., *Anal. Chem.* **24**, 674 (1952).
24. Small, L. A., *Analyst* **84**, 117 (1959).
25. Jordan, D. E., *Anal. Chim. Acta* **30**, 297 (1964).
26. Baumann, W. M. and Simon, W., *Z. anal. Chem.* **216**, 273 (1966).
27. Fritz, J. S. and Schenk, G. H., *Anal. Chem.* **31**, 1808 (1959).
28. Schenk, G. H. and Fritz, J. S., *Anal. Chem.* **32**, 987 (1960).
29. Klouček, B., Gasparič, J. and Obruba, K., *Collection Czech. Chem. Commun.* **28**, 1606 (1963).
30. Magnuson, J. A. and Cerri, R. J., *Anal. Chem.* **38**, 1088 (1966).
31. Jordan, D. E., *J. Am. Oil Chem. Soc.* **41**, 500 (1964); *Chem. Abs.* **61**, 6399 (1964).
32. Madeline, P. and Debourg, G., *Mem. Poudres* **43**, 143 (1961); *Chem. Abs.* **58**, 10731 (1963).
33. Stetzler, R. S. and Smullin, C. F., *Anal. Chem.* **34**, 194 (1962).
34. Pietrzyk, D. J. and Belisle, J., *Anal. Chem.* **38**, 1508 (1966).

35. VERLEY, A. and BÖLSING, F., *Ber.* **34**, 3354 (1901).
36. DELABY, R. and SABETAY, S., *Bull. Soc. Chim. France* [5] **2**, 1716 (1945).
37. HAUSCHILD, R., SINGER, K. and PETIT, J., *Bull. Soc. Chim. France* 768 (1956).
38. QUATTRONE, JR., J. J. and CHOY, T., *Microchem. J.* **6**, 259 (1962).
39. BENSON, R. H. and TURNER, R. B., *Anal. Chem.* **32**, 1464 (1960).
40. KARTHA, A. R. S. and KIDWAI, M. A., *Indian J. Chem.* **3**, 371 (1965); *Chem. Abs.* **63**, 16621 (1965).
41. SMITH, D. M. and BRYANT, W. M. D., *J. Am. Chem. Soc.* **57**, 61 (1935).
42. PESEZ, M., *Bull. Soc. Chim. France* 1237 (1954).
43. SULLY, B. D., *Analyst* **87**, 940 (1962).
44. ROBINSON, JR., W. T., CUNDIFF, R. H. and MARKUNAS, P. C., *Anal. Chem.* **33**, 1030 (1961).
45. ROBINSON, JR., W. T., CUNDIFF, R. H., SENSABAUGH, A. J. and MARKUNAS, P. C., *Talanta* **3**, 307 (1960).
46. SMITH, B. and HAGLUND, Å., *Acta Chem. Scand.* **15**, 675 (1961).
47. SIGGIA, S., HANNA, J. G. and CULMO, R., *Anal. Chem.* **33**, 900 (1961).
48. KINGSTON, B. H. M., GAREY, J. J. and HELLWIG, W. B., *Anal. Chem.* **41**, 86 (1969).
48a. OPPENAUER, R. V., *Monatsh. Chem.* **97**, 67 (1966).
48b. MESNARD, P., GIBIRILA, B. and BERTUCAT, M., *Chim. Anal. (Paris)* **45**, 491 (1963).
48c. MESNARD, P., GIBIRILA, B. and BERTUCAT, M., *Kongr. Pharm-Wiss. Vortr., Originalmitt. Münster* **23**, 603 (1963); *Chem. Abs.* **62**, 6335 (1965).
48d. IYER, V. and MATHUR, N. K., *Anal. Chim. Acta* **33**, 554 (1965).
48e. NARANG, C. K., IYER, V. and MATHUR, N. K., *Microchem. J.* **9**, 408 (1965).
49. SCHENK, G. H. and SANTIAGO, M., *Anal. Chem.* **39**, 1795 (1967).
50. DREHER, B., *Farbe und Lack* **67**, 703 (1961); *Chem. Abs.* **56**, 6116 (1962).
51. REED, D. H., CRITCHFIELD, F. E. and ELDER, D. K., *Anal. Chem.* **35**, 571 (1963).
52. OTEY, F. H., ZAGOREN, B. L. and MEHLTRETTER, C. L., *J. appl. Polym. Sci.* **8**, 1985 (1964); *Chem. Abs.* **62**, 1745 (1965).
53. VEIBEL, S., *Ber.* **63**, 1582 (1930).
54. SMITH, B., *Acta Chem. Scand.* **10**, 1589 (1956); **11**, 839 (1957).
55. CHIKRYZOVA, E. G. and KHOMENKO, V. A., *Zh. Anal. Khim.* **23**, 1206 (1968); *Chem. Abs.* **70**, ref. 8685 (1969).
56. SCHOENBERGER, E., *Studia Univ. Babes-Bolyai, Ser. Chim.* **11**, 146 (1966); *Chem. Abs.* **66**, ref. 121902 (1967).
57. KARPOV, O. N., *Zh. Anal. Khim.* **23**, 1423 (1968); *Chem. Abs.* **70**, ref. 8681 (1969).
58. HUBER, C. O. and GILBERT, J. M., *Anal. Chem.* **34**, 247 (1962).
59. KONISHI, K., MORI, Y., INOUE, H. and NOZOE, M., *Anal. Chem.* **40**, 2198 (1968).
60. BRADLEY, M. P. T. and PENKETH, G. E., *Analyst* **92**, 701 (1967).
61. CHENG, F. W., *Anal. Lett.* **1**, 463 (1968); *Chem. Abs.* **69**, ref. 32845 (1968).
62. JAULMES, P. and MESTRES, R., *Chim. Anal.* **40**, 413 (1958).
63. PANWAR, K. S. and GAUR, J. N., *Anal. Chim. Acta* **33**, 318 (1965).
64. GUILBAULT, G. G. and MCCURDY, JR., W. H., *Anal. Chem.* **33**, 580 (1961).
65. PESEZ, M. and BARTOS, J., *Ann. Pharm. France* **22**, 609 (1964).
66. SRIVASTAVA, V. N. P. and SAXENA, O. C., *Microchem. J.* **12**, 435 (1967).
67. JASELSKIS, B. and WARRINER, J. P., *Anal. Chem.* **38**, 563 (1966).
68. WILLIAMSON, S. M. and KOCH, C. W., *Science* **139**, 1046 (1963).
69. BLOIS, M. S., *Nature* **181**, 1199 (1958).
70. PAPARIELLO, G. J. and JANISH, M. A. M., *Anal. Chem.* **37**, 899 (1965); **38**, 211 (1966).
71. SCHENK, G. H. and BROWN, D. J., *Talanta* **14**, 257 (1967).
72. BRYANT, W. M. D., MITCHELL, JR., J. and SMITH, D. M., *J. Am. Chem. Soc.* **62**, 1 (1940).

73. Petrova, L. N. and Novikova, E. N., *Zh. Anal. Khim.* **12**, 411 (1957); *J. Anal. Chem. USSR* **12**, 429 (1957); *Chem. Abs.* **52**, 1860 (1958).
74. Zerevitinov, T., *Ber.* **41**, 2233 (1908).
75. Zaugg, H. E. and Lauer, W. M., *Anal. Chem.* **20**, 1022 (1948).
76. Terent'ev, A. P. and Kirejeva, A., *J. Isvest. Akad. Nauk SSSR, Otdel. Khim. Nauk* 172 (1951); *Chem. Abs.* **45**, 6964 (1951).
77. Stenmark, G. A. and Weiss, F. T., *Anal. Chem.* **28**, 1784 (1956).
78. Falgoux, D., Demeure, M., Engel, J. and Granger, C., *Bull. Soc. Chim. France* 2516 (1967).
79. Norton, E. J., Turner, L. and Salmon, D. G., *Analyst* **95**, 80 (1970).
80. Chumachenko, M. N., Tverdynkova, L. B. and Leenson, G. F., *Zh. Anal. Khim.* **21**, 617 (1966); *J. Anal. Chem. USSR* **21**, 550 (1966); *Chem. Abs.* **65**, 9709 (1966).
81. Koelsch, C. F., *J. Am. Chem. Soc.* **53**, 304 (1931).
82. Schotte, L. and Veibel, S., *Acta Chem. Scand.* **7**, 1357 (1953).
83. Jureček, M., Chládek, O., Chládková, R., Souček, M. and Srpová, B., *Collection Czech. Chem. Commun.* **22**, 1809 (1957).
84. Berger, J., *Acta Chem. Scand.* **6**, 1564 (1952).
85. Whitmore, W. F. and Lieber, E., *Ind. Eng. Chem. Anal.* **7**, 127 (1935).
86. Kolthoff, I. M. and Robinson, C., *Rec. Trav. Chim. Pays-Bas* **45**, 169 (1926).
87. Vodzinskii, Yu. V. and Semchikova, C. S., *Tr. po Khim. i Khim.-Technol.* 272 (1963); *Chem. Abs.* **61**, 6640 (1964).
88. Fürst, W., *Pharm. Zentralh.* **107**, 184 (1968); *Chem. Abs.* **69**, ref. 5227 (1968).
89. Silkina, R. P., Zhuk, M. A. and Kirilyus, I. V., *Tr. Khim.-Met. Inst. Akad. Nauk. Kaz. SSR* 136 (1968); *Chem. Abs.* **70**, ref. 8683 (1969).
90. Tanaka, M., *Talanta* **5**, 162 (1960).
91. Pesez, M. and Bartos, J., *Bull. Soc. Chim. France* 1930 (1961).
92. Stiller, M., *Anal. Chim. Acta* **25**, 85 (1961).
93. Bartos, J., *Ann. Pharm. Fr.* **27**, 323 (1969).
94. Jacob, D. R., *Anal. Chim. Acta* **44**, 449 (1969).
95. Maros, L., Perl, I. and Szakacs-Pinter, M., *Ann. Univ. Sci. Budapest Rolando Eötvös Nominatae, Sect. Chim.* **7**, 37 (1965); *Chem. Abs.* **65**, 8008 (1966).
96. Omboly, C. and Derzsi, E., *Acta Pharm. Hung.* **32**, 246 (1962); *Chem. Abs.* **58**, 9634 (1963).
97. Hubicka, K., *Dissertationes Pharm.* (*Warszawa*) **15**, 453 (1963); *Chem. Abs.* **61**, 11340 (1964).
98. Mizukami, S. and Nagata, K., *Yakugaku Zasshi* **81**, 1285 (1961); *Chem. Abs.* **56**, 8005 (1962).
99. Gutnikov, G. and Schenk, G. H., *Anal. Chem.* **34**, 1316 (1962).
100. Obtemperanskaya, S. I. and Kosina, A. N., *Vestn. Mosk. Univ. Ser. II, Khim.* **19**, 80 (1964); *Chem. Abs.* **61**, 10046 (1964).
101. Obtemperanskaya, S. I. and Dyk-Hoe, N., *Vestn. Mosk. Univ. Ser. II, Khim.* **19**, 83 (1964); *Chem. Abs.* **61**, 10047 (1964).
102. Pesez, M. and Bartos, J., *Talanta* **14**, 1097 (1967).
103. McKinney, R. W. and Reynolds, C. A., *Talanta* **1**, 46 (1958).
104. Scoggins, M. W., *Anal. Chem.* **36**, 1152 (1964).
105. Pesez, M. and Bartos, J., *Talanta* **5**, 216 (1960).
106. Obtemperanskaya, S. I. and Borodina, G. P., *Vestn. Mosk. Univ. Ser. II, Khim.* **22**, 88 (1967); *Chem. Abs.* **66**, ref. 121886 (1967).
107. Janovsky, J. V., *Ber.* **24**, 971 (1894).
108. Shmulyakovskii, Y. E., *Zhur. Priklad. Khim.* **32**, 2513 (1959); *Chem. Abs.* **54**, 7437 (1960).

109. Meluzova, G. B., Kotel'nikov, B. P. and Prokhorova, Z. A., *Zh. Anal. Khim* **17**, 362 (1962); *Chem. Abs.* **57**, 7403 (1962).
110. Crummett, W. B., *Anal. Chem.* **34**, 1147 (1962).
111. Katz, I. and Keeney, M., *Anal. Chem.* **36**, 231 (1964).
112. Popa, I., Voicu, A., Armasescu, T., Stan, T. and Ioanid, N., *Farmacia (Bucharest)* **13**, 287 (1965); *Chem. Abs.* **63**, 6792 (1965).
113. Hoare, D. E., Ogilvie, R. R. and Wellington, C. A., *Anal. Chem.* **38**, 1799 (1966).
114. Etlinger, M. B., Ruchhoft, C. C. and Lishka, R. J., *Anal. Chem.* **23**, 1783 (1951).
115. Shaw, J. A., *Anal. Chem.* **23**, 1788 (1951).
116. Emerson, E., *J. Org. Chem.* **8**, 417 (1943).
117. Emerson, E. and Kelly, K., *J. Org. Chem.* **13**, 532 (1948).
118. Goddu, R. F., *Anal. Chem.* **30**, 2009 (1958).
119. Hilton, C. L., *Anal. Chem.* **31**, 1610 (1959).
120. Kyriacou, D., *Anal. Chem.* **33**, 153 (1961).
121. Anderson, D. M. W. and Zaidi, S. S. H., *Talanta* **10**, 691 (1963).
122. Scoggins, M. W. and Miller, J. W., *Anal. Chem.* **38**, 612 (1966).
123. Nisli, G. and Townshend, A., *Talanta* **15**, 1377 (1968).
124. Keith, R. W., Le Tourneau, D. and Mahlum, D., *J. Chromatog.* **1**, 534 (1958).
125. Kaufmann, H. P. and Kessen, G., *Z. physiol. Chem.* **317**, 43 (1959).
126. Cerutti, G. and Wezzini, W., *Chim. Ind. (Milano)* **43**, 784 (1961); *Chem. Abs.* **60**, 13890 (1964).
127. Vasyunina, N. A., Balandin, A. A., Mamatov, Yu. and Pustovaya, L. M., *Gidroliz i Lesokhim. Prom.* **15**, 13 (1962); *Chem. Abs.* **57**, 1558 (1962).
128. Gasparič, J. and Gemzova, I., *J. Chromatog.* **35**, 362 (1968).
129. Bate-Smith, E. C. and Westall, R. G., *Biochim. Biophys. Acta* **4**, 427 (1950).
130. Kuffner, F. and Kallina, D., *Monatsh. Chem.* **90**, 463 (1959).
131. Kallina, D. and Kuffner, F., *Monatsh. Chem.* **91**, 289 (1960).
132. Sokolov, V. A. and Kolesnikova, L. P., *Neftekhimiya* **1**, 564 (1961); *Chem. Abs.* **56**, 14904 (1962).
133. Hofmann, M. and Struppe, H. G., *Acta Chim. Acad. Sci. Hung.* **27**, 239 (1961).
134. Robinson, J. W., *Anal. Chim. Acta* **27**, 377 (1962).
135. Rudol'fi, T. A., Shchedrina, M. M. and Kokhmanskii, A. V., *Tr. Vses. Nauchn-Issled Inst. Sintetich i Natural'n. Dushistykh Veshchestv.* **6**, 104 (1963); *Chem. Abs.* **61**, 12639 (1964).
136. Ikeda, R. M., Simmons, D. E. and Grossman, J. D., *Anal. Chem.* **36**, 2188 (1964).
137. Schlunegger, U. P., *J. Chromatog.* **24**, 165 (1966).
138. Bluestein, C. and Posmanter, H. N., *Anal. Chem.* **38**, 1865 (1966).
139. Doelle, H. W., *J. Gas Chromatog.* **5**, 582 (1967).
140. Schwartz, R. D. and Mathews, R. G., *J. Chromatog. Sci.* **7**, 593 (1969).
141. Minyard, J. P., Tumlinson, J. H., Thompson, A. C. and Hedin, P. A., *J. Chromatog.* **29**, 88 (1967).
142. Kolesnikova, L. P., Khotimskaya, M. J., Gurevich, V. T., Martynes, M., Kamzolkin, V. V. and Bashkirov, A. N., *Neftekhimiya* **8**, 722 (1968); *Chem. Abs.* **70**, ref. 34094 (1969).
143. Kotel'nikov, B. P. and Nichikova, P. R., *Khim. Tekhnol. Topl. Masel* **15**, 32 (1970); *Chem. Abs.* **72**, ref. 139429 (1970).
144. Balakhontseva, V. N. and Poltinina, R. M., *Zh. Anal. Khim.* **19**, 757 (1964); *Chem. Abs.* **61**, 11339 (1964).
145. Weatherall, I. L., *J. Chromatog.* **26**, 251 (1967).
146. Lindsay Smith, J. R. and Waddington, D. J., *J. Chromatog.* **36**, 145 (1968).

147. Novoselov, A. I., Afnas'ev, A. M., Kalyazin, E. P. and Zakharov, V. F., *Zh. Anal. Kh.* **25**, 386 (1970); *Chem. Abs.* **72**, ref. 139428 (1970).
148. Bergman, G. and Jentzsch, D., *Z. anal. Chem.* **164**, 10 (1958).
149. Janák, J. and Komers, R., *Z. anal. Chem.* **164**, 69 (1958).
150. Janák, J., Komers, R. and Šima, J., *Collection Czech. Chem. Commun.* **24**, 1492 (1959).
151. Janák, J. and Komers, R., *Collection Czech. Chem. Commun.* **24**, 1960 (1959).
152. Fitzgerald, J. S., *Australian J. Appl. Sci.* **10**, 169 (1959); *Chem. Abs.* **53**, 14844 (1959).
153. Méndez, J. and Stevenson, F. J., *J. Gas Chromatog.* **4**, 483 (1966).
154. Méndez, J., *J. Gas Chromatog.* **6**, 168 (1968).
155. Landault, C. and Guiochon, G., *Anal. Chem.* **39**, 713 (1967).
156. Clifford, D. R. and Watkins, D. A. M., *J. Gas Chromatog.* **6**, 191 (1968).
157. Malz, F. and Gorlas, J., *Z. anal. Chem.* **242**, 81 (1968).
158. Delaunois, C. and Bettens, B., *Ann. Mines Belg.* 633 (1968); *Chem. Abs.* **70**, ref. 64162 (1969).
159. Nosal, A., Pasternak, A. and Witek, S., *Chem. Anal.* (*Warszawa*) **14**, 1115 (1969); *Chem. Abs.* **72**, ref. 128478 (1970).
160. Zakupra, V. A. and Dobrov, V. S., *Khim. Tekhnol. Topl. Masel* **14**, 54 (1969); *Chem. Abs.* **72**, ref. 50757 (1970).
161. Sato, M., Miyake, H., Mitooka, M. and Asano, K., *Bull. Soc. Chim. Japan* **38**, 884 (1965).
162. Larkham, T. W. and Pagington, J. S., *J. Chromatog.* **28**, 422 (1967).
163. Wood, R., *J. Gas Chromatog.* **6**, 94 (1968).
164. Neurath, G. and Lüttich, W., *J. Chromatog.* **34**, 253 (1968).
165. Volodina, M. A. and Kon'kova, I. V., *Vestn. Mosk. Univ. Khim.* **11**, 119 (1970); *Chem. Abs.* **73**, ref. 10500 (1970).
166. Galetto, W. G., Kepner, R. E. and Webb, A. D., *Anal. Chem.* **38**, 34 (1966).
167. Bhattacharyya, A. C., Bhattacharjee, A., Guha, O. K. and Basu, A. N., *Anal. Chem.* **40**, 1873 (1968).
168. Cohen, I. C., Norcup, J., Ruzicka, J. H. A. and Wheals, B. B., *J. Chromatog.* **44**, 251 (1969).
169. Cohen, I. C. and Wheals, B. B., *J. Chromatog.* **43**, 233 (1969).
170. Langer, S. H., Pantages, P. and Wender, I., *Chem. Ind.* (*London*) 1664 (1958).
171. Freedman, R. W. and Croitoru, P. P., *Anal. Chem.* **36**, 1389 (1964).
172. Sweeley, C. C., Bentley, R., Makita, M. and Wells, W. W., *J. Am. Chem. Soc.* **85**, 2497 (1963).
173. Leibman, K. C. and Ortiz, E., *J. Chromatog.* **32**, 757 (1968).
174. Dutton, G. G. S., Gibney, K. B., Jensen, G. D. and Reid, P. E., *J. Chromatog.* **36**, 152 (1968).
175. van Ling, G., *J. Chromatog.* **44**, 175 (1969).
176. Kushner, I., Barr, P. A. and Chortyk, O. T., *Anal. Chem.* **42**, 1619 (1970).
177. Zahn, C., Sharkey, Jr., A. G. and Wender, I., *U.S. Bur. Mines, Rept. Invest. No. 5976*, 33 pp. (1962); *Chem. Abs.* **57**, 4038 (1962).
178. Mathias, A., *Anal. Chim. Acta* **31**, 598 (1964).
179. Chapman, O. L. and King, R. W., *J. Am. Chem. Soc.* **86**, 1256 (1964).
180. Godsey, C. E., *Anal. Chem.* **38**, 842 (1966).
181. Babiec, Jr., J. S., Barrante, J. R. and Vickers, G. D., *Anal. Chem.* **40**, 610 (1968).
182. Siggia, S., *Quantitative Organic Analysis via Functional Groups*, 3rd Edn., p. 528, Wiley & Sons, New York (1963).

# AUTHOR INDEX

Numbers in *italics* indicate the pages on which references are listed in full.

# SUBJECT INDEX